THE ELEARNING & INSTRUCTIONAL DESIGN ROADMAP

AN UN-BORING GUIDE FOR NEWBIES, CAREER-CHANGERS, AND ANYONE WHO WANTS TO BUILD BETTER ELEARNING

AUBREY COOK

ODDLY SHARP LLC

Cover & Book Designer: Aubrey Cook
Editor: Chandi Lyn

ISBN (paperback): 979-8-9887622-0-1
ISBN (kindle ebook): 979-8-9887622-1-8

Published by Oddly Sharp LLC

CONTENTS

For my parents. Something different.

INTRODUCTION

In hindsight, you're not sure how such a promising job went south so quickly. But there you were, shouldering a project that teetered between hot mess and full-on train wreck.

It began three weeks ago when your employer asked you to put together training for new hires. This was a first for you, but with a background in education and a few design courses under your belt, why not? You'd suss out the details along the way!

So, you dove in, gathering material from department leads and organizing it into (what you hoped was) a logical order. You wrote a script and produced homegrown media to break things up. But as the content swelled and choices multiplied, you felt less and less sure of the next steps.

Hungry for guidance, you scoured every available resource, pouring through articles and videos, even dusting off some Very Official-Looking textbooks on instructional design. Amazed at the depth of knowledge but shocked by a lack of practical advice, you grew frustrated and abandoned the unhelpful academic jargon.

As the release date approached, you stepped back to review your work, and well, all training was supposed to be a little dry, right? Sure, some parts felt disjointed; maybe inconsistent. What mattered was that you'd mixed all the ingredients for your learners to consume, like a nourishing, educational slurry.

But no. This slurry didn't go down so smoothly. And your intended audience found it neither delicious nor satisfying.

People came away with more questions than answers, skimming content with half-glazed expressions and the dreaded eye roll. Even more discouraging, your role had been to help people with crucial aspects of their jobs, but in the following weeks, it became clear they were confused, frustrated, and—worse—totally unprepared for what they needed to do. And while you knew the experience was lacking, you had no idea what went wrong or where to look for direction.

Well... crud.

Most of us have experienced something like this from one side or the other. Because anyone who's had a job or taken courses online is familiar with training of some kind. And the more experience we have, the more familiar we are with *awful* training—either because it's ineffective or just plain excruciating.

Still, there's no avoiding it. Whether you employ ten or ten thousand people, you'll need education on various topics, from workplace safety to career development and everything in between.

Gone are the days when instruction could be rubber-stamped with a one-size-fits-all video, complete with cringe-worthy dialogue and fashion to match.

Delivery mechanisms have become sophisticated and interactive. Analytics and measurable results are now prerequisites. And as learning experiences grow more nuanced, the people who build them need a proven approach to guide their work.

While instructional design offers plenty of guidance, there's a huge gap between academic theory and on-the-job tactics. Sources can be dense, and distilling a step-by-step roadmap from all the models, principles, and frameworks can feel like sifting for gold in a waterfall.

But what if you already had a workflow blocked out and ready to roll? You could just sit down to a project, confident in your steps and the tools at your disposal. You'd be free to focus on problem-solving rather than defining a process.

My two cents? Avoid paving the road while you're driving on it.

What will you learn?

Whether you're a new instructional designer, looking to transition from your current job, or tasked with training for your organization, this book will serve as a how-to guide. I've distilled some of the most useful tips, tactics, and research-backed learning principles and arranged them into an actionable workflow.

We'll cover topics like:

- Understanding eLearning and instructional design
- How memory works and ways to design for sticky learning

- Professional tools, models, and frameworks
- Working with clients and subject matter experts
- Defining your learners and goals
- Outlines, storyboards, and visual design
- Style guides and layout templates
- Creating and presenting prototypes
- Adding interactivity and multimedia
- Implementing and evaluating your learning experience

By the end of this book, you should have a foundation to design and launch an eLearning project from start to finish. So whether you're looking for help with your career, tactical tips for a learning project, or are just curious about the field, let's get rolling!

Why did I write this book?

I spent years mentoring people in visual design, UX, and web development. Along the way, I found that some concepts stuck instantly, others took time and practice, and still others bounced clean off the brain like frogs from a trampoline. Over time, I became fascinated not only by design systems and methods of communication, but with how people absorb information. I knew there had to be a systematic way to tackle such things. So, after much reading, research, and distilling wisdom from brains much bigger than my own, here we are! I hope you find it useful.

A software reality

Any software or technology mentioned in this book will eventually be outdated or replaced with newer, shinier offerings. While I hesitate to call out specific products, some of the most common questions about instructional design relate to digital tools.

With that in mind, I've included a semi-quarantined *Tools & Resources* section at the back with links to software options, design utilities, and other useful tech. You can use these as starting points before choosing your favorites. Approach the process with a willingness to adapt, and always keep an eye out for new tools.

Also, I have no affiliation with any software mentioned herein, nor do I represent the developers or parent companies in any way. Experiment, use your discretion, and disregard them at will.

1

INSTRUCTIONAL DESIGN 101

At its most basic, *instructional design* (ID) is about developing content to help others learn. You'll find it everywhere—education, corporate training, non-profit organizations, and even the gaming world! It provides a framework to guide you through each stage of learning design, from drafting project goals to measuring results.

Companies often hire *instructional designers* to improve products and address performance issues. While they may hold alternate titles like course developer or training design specialist, their goal is always the same: to design effective learning experiences.

What does it take to be an instructional designer?

Above all, instructional designers need an understanding of how people learn. They consider the mechanics of memory, how to make information more sticky, and ways to engage learners using the latest, relevant tech.

Many people wind up in the role after it catches their eye from an adjacent field or because they suddenly become responsible for training. Teachers are frequent converts, as are graphic and user experience designers.

In addition to ID itself, instructional designers often major in education or psychology. Some positions may require a master's or higher, but specific degrees aren't necessarily barriers to employment. Most important is the strength of your portfolio (which, if you've got some grit, you can build without work experience) and an ability to use standard practices.

Long story short, if you're passionate about the field but lack a related degree, don't let it discourage you. Just research requirements for the jobs you want, develop your expertise, and be ready to adapt!

Tools of the trade

Any instructional designer worth their salt needs a working knowledge of industry methods. There's no waffling on this one.

Fortunately, the resources at your disposal are vast. Need to figure out project goals? Check! Looking for ways to organize your workflow? We've got at least four! And don't get me started on all the tools for storyboarding and evaluating your build.

In addition to conceptual chops, instructional designers should have experience with authoring tools, learning management systems, and digital media. They need strong planning skills and the ability to guide everyone involved—subject matter experts, clients, and developers.

Sounds like a lot, right? But by the end of this book, you'll have a firm grip on the fundamentals and enough confidence to explore on your own.

And here's a tip to ease some anxiety: don't worry too much about memorization. If you're in mid-project one day and say: "I can't recall my storyboard fields, but I know there's a guide in book X." Well then, you understand what you need to do, you know there's a technique you can use, and you have a pretty good idea of where to find it.

And that's perfectly fine, in my opinion.

Why focus on eLearning?

Online video courses, interactive tutorials, and mobile apps that teach your dog to make spanakopita—all eLearning. If it's digital and teaches you something new, it probably falls into this category.

'Nuff said.

Some types of eLearning happen in real-time and allow participants to interact with each other and the instructors. Others are asynchronous and move at an individual pace.

Why does this book focus mostly on eLearning techniques? I'll give you three reasons.

First, instructional design is used for all kinds of learning experiences, not just those in the digital realm. Some principles are universal; others vary by delivery. An approach for instructor-led, in-person training might prove irrelevant when designing self-guided, online courses.

While eLearning won't suit every situation, focusing on one type of instruction narrows the scope of this book. It lets us dive into eLearning tactics rather than leapfrogging between in-person and digital methods.

Second, eLearning leans heavily on instructional design principles because there's less guidance from an in-person instructor. Most eLearning needs to be effective all by its lonesome.

Finally, it's no surprise that distance learning has become indispensable. Most organizations have remote workers who need the same training as those on-site. But how do you address such things when your staff of thousands is scattered across the country or the world?

However you slice it, eLearning is expanding, and there's a growing need for creators. It's a perfect time to roll up your sleeves and get in there!

Instructional designer vs. eLearning developer

Let's talk about the people who design and build eLearning experiences.

If we're going by traditional definitions, you can think of instructional designers as the architects of a learning experience and *eLearning developers* as the hands-on builders.

An instructional designer often guides the first phase of a learning experience. They're responsible for defining goals, creating a narrative for the learner, and outlining a general blueprint. They also devise ways to evaluate the experience after it's live—how are learners performing, and how might results be improved?

Their job may involve:

- Working with stakeholders and subject matter experts

- Defining what learners need to accomplish via learning objectives or action maps
- Creating scenarios, scripts, and storyboards
- Visual design
- Planning practice activities and interactivity
- Evaluating results

After the instructional designer creates a blueprint, an eLearning developer is often responsible for technical execution and tasked with things like:

- Prototyping
- Building or coding the experience using artifacts from the instructional designer
- Production design (a.k.a. creating and optimizing visual assets like images or video)

However, sometimes the instructional designer assumes both roles, depending on the staff size and project complexity.

In this book, we'll focus on skills for a multi-disciplinary instructional designer—someone who can assess, plan, and outline a learning experience, and delve into eLearning developer tasks like production design and prototyping.

Requirements for every job differ, but the broader your skill set, the more valuable you are.

Related fields and their goodies

Remember that family reunion where nothing was awkward, and everyone kept their idiosyncrasies in check? Me neither! But sometimes, relatives can be a source of support and guidance—or, at the very least, excellent hand-me-downs.

As you dive into instructional design, you'll find fields with similar (and occasionally confusing) titles and terms. Often, they're closely related to ID and may even be used interchangeably by companies and hiring managers. Some of them draw from shared fundamentals, and it helps to understand what they offer.

Once you know the basics, you'll know where to look when you need to borrow a method (e.g., wireframing from UX or design principles from visual design). You'll also understand which roles should be responsible for each step in a team setting.

We won't be taking an extended tour here. I'll just give you the broad strokes so you can borrow as needed.

User experience (UX) design

User experience (UX) *designers* analyze how people interact with websites and apps, creating products or services with the user's needs in mind. They collect research, analytics, and testing data to define requirements and pain points. They use these insights to design intuitive interfaces.

As you'll see below, several fields overlap with UX design. And as an instructional designer, you'll borrow techniques from all of them.

Information architecture (IA)

It may be easiest to think of *information architecture* (IA) as one piece of the user experience design puzzle. If UX design focuses on the big picture of a usable experience, information architecture defines a more tactical process.

Among other things, IA examines how information appears in a digital experience, how user interfaces (UI) function, and how users navigate an app or website.

UX designers are versed in information architecture and will either create their own IA deliverables (wireframes, sitemaps, etc.) or work directly with *information architects.*

Instructional designers borrow several tactics from UX and IA. They may create *personas* (example users with shared characteristics), *wireframes* (un-styled layouts of user interfaces), *user flows* (paths users can take as they move through an experience), and *sitemaps* (an overview of all pages or states).

There's lots of good stuff from the UX and IA disciplines.

Visual design (a.k.a. graphic design)

Simply put, a *visual designer* (usually called graphic designer or just designer) is tasked with making things look good. Whether it's a website, an app, a brochure, or a presentation, these specialists create content that's aesthetically pleasing and on-brand. But it's not just about making things look pretty. Visual designers also consider function and meaning.

While there's plenty of room for personal style, there are established principles for good design. Things like proper alignment and contrast make it easier for the eye and brain to consume information.

When working in the digital space, these designers also consider usability. A stunning site is useless if no one can navigate its content or find what they need. Visual designers balance form and function to create effective results.

Does this sound like it's starting to bleed into UX design? Indeed! Visual designers often inherit wireframes from their

UX counterparts. They apply color, type, spacing, and other aesthetic goodies to bring black-and-white layouts to life in their *comps*. You can think of it as putting clothes on a mannequin or, in the more grizzly but common vernacular, skinning the wireframes.

We'll be borrowing quite a bit from visual design in this book. At a minimum, instructional designers should master design principles, style guides, and grids.

And yes, I'll give you a solid foundation for all that.

Learning experience design

Distinguishing between instructional design and *learning experience design* (LXD) via internet searches may give you a headache. I wouldn't spend too much time unraveling this twisted noodle at this point in your career, so I'll keep it short.

A few key elements go into LXD: understanding your learners, focusing on their goals, and harnessing other disciplines like UX design, interaction design, and visual design. However, if you only went by these traits, you'd be hard-pressed to find an instructional designer who doesn't consider the same things in their job.

According to Niels Floor, who coined the term in 2007, "ID has a more scientific perspective as an applied science while LXD has a more creative perspective as an applied art" (Floor, 2021).

While some consider these differences fairly nuanced, it often comes down to opinion or specific organizational roles.

In short, if you see a job posting for Learning Experience Designer and are confident in your instructional design skills, take a peek at the requirements. There's a good chance your skills will translate.

And along those lines...

If you find yourself overwhelmed by acronyms, initialisms, or fancy job titles along the way, step back and ask:

- What tasks are they doing?
- What parts of a project are they responsible for?
- What tangible things are they creating?
- Are they using methods that sound useful for my next project?

The moral of this story: borrow everything

The boundaries between ID-adjacent fields can be fuzzy, and in practice, roles may overlap or vary between companies.

Many of these fields share common ground and address similar problems. So if you find a neighboring technique that sounds useful for your ID project, don't be afraid to toss it in the mix. Super glue was used to patch wounds during the Vietnam War, but it's also great for shattered pottery.

We'll be doing plenty of discipline borrowing in later chapters, especially from visual and user experience design. So warm up those fast-moving fingers and get ready to swipe!

Chapter takeaways

And we're off! We've introduced industry roles, requirements, and relatives worth investigating. Remember the following tidbits as we move forward, and in the next chapter, we'll put memory under the microscope.

- *Instructional design (ID)* is a system for creating materials to help people learn more effectively.

- *Instructional designers* come from various backgrounds. They may pursue dedicated degrees, become responsible for training while working in other roles, or transition from related fields like education.
- While a related degree may be optional, instructional designers require a working knowledge of ID practices. They should also be well versed in UX and visual design.
- eLearning defines a range of digital experiences, including online lessons, video instruction, and distance-learning classes.
- Traditionally, instructional designers create the blueprint for an eLearning experience while eLearning developers handle the technical build. However, in some organizations, the instructional designer may assume both roles.
- Many fields related to instructional design have similar or overlapping principles. It's helpful to understand the differences and commonalities so you can borrow techniques.
- The more cross-discipline knowledge you have, the more valuable you are to your current organization and as a potential hire.

2

DESIGNING FOR MEMORY

Wouldn't it be great if we could absorb information effortlessly? We'd never worry about forgetting state capitals, how to say "Who goes there?" in Swedish, or the main spice in paella.

Turmeric? Maybe Saffron?

Fortunately, instructional designers have plenty of ways to help people recall information more easily. Not only do these methods aid in memorization, but they form deeper connections with the material. There's a big difference between parroting ideas and applying them creatively.

Don't worry. We won't spend half this book delving into neuroscience. You're here for practical, on-the-job methods, and by the glutes of Thor, that's what you'll get. But knowing how memory works is essential for instructional designers.

Speaking of glutes (yes, I actually wrote that), building strong memories is similar to developing muscle. Some methods work while others don't, and retention demands energy in the form of mental focus. Memories respond to variety and repetition over time, and there's no one-size-fits-all solution.

When designing a learning experience, you want to stimulate learners, choose appropriate techniques to match the material, and, occasionally, wake them up so they pay attention and push memories into deeper storage.

Yeah. I'm pretty sure it's saffron.

Different memory systems

It's no secret that memory and learning are extremely dense topics. People spend their lives working in this field and only arrive at more questions. For now, I'll do my best to distill this into practical nuggets we can use and discuss. Just keep in mind that as research continues, this knowledge may change.

Humble disclaimers aside, the four memory systems include:

- *Sensory*
- *Short-term*
- *Working*
- *Long-term*

Rather than working separately, studies suggest that these systems connect in a progression, with one stage leading to the next.

You could think about *sensory memory* as your driveway (the beginning of a trip), *short-term* as the path leading out of your neighborhood, *working* as your city streets, and *long-term* as the freeway, where your brain goes on cruise control because your data is firmly seated.

If someone learns something but only uses it for a short time—such as once in a sentence—it won't make its way through the rest of the path and into long-term memory.

However, if someone hears a word (sensory), uses it in a sentence (short-term), is asked the meaning of it the next day (working), and then continues to use it every day for a week (long-term), there is a good chance they'll remember it permanently, or rather, that their brain will file it away for later retrieval.

The more you reinforce a memory at each stage, the more likely that information is to make it all the way into long-term storage—much like recharging your car so it has enough power to reach the next stop.

Sensory memory

Sensory memory holds information for a brief moment, often less than a second. Once your brain acknowledges it, it begins to move into a deeper memory pathway. Your brain can't process every stimulus from the surrounding world, so it filters out the rubbish and only handles things that grab your focus.

Short-term memory

Short-term memory lasts about thirty seconds and can hold roughly seven items at once, give or take.

If we picture these memory stages as a progression, items move from sensory memory into short-term. As new memories slide in, the old ones either get enough attention to advance, or they're tossed.

You use short-term memory to remember a phone number while searching for a pen, or to hold last-minute grocery requests as you enter a store.

Working memory

Working memory helps you retain information about what you're doing right now. It gives you space to manipulate data for the task at hand, such as solving multi-step math problems, baking a quiche and doubling the ingredients, or processing another person's argument and devising a response.

Short-term and working memory are distinct, even though their roles overlap significantly.

Long-term memory

When you want information to stick around for a while, make sure it gets to long-term memory! The phrase long-term isn't necessarily what you'd expect, though. Any information that survives short-term and working memory begins to classify as long-term.

Long-term memories are subdivided into *implicit* and *explicit* buckets.

Implicit memories help make up who you are—things that form your persona without conscious thought. These include procedures you can perform on auto-pilot, like driving or brushing your teeth. This is also known as *automatic memory*.

We can break down explicit memories into *episodic* and *semantic* types. Episodic memories hold pictures and feelings of past events and facts about your life. On the other hand, semantic memories contain general knowledge, like the fact that mammals breathe air and sea otters hold hands while they sleep. Adorable!

Targeting memory systems in eLearning

Okay, let's put this into practice.

Again, an instructional designer aims to help people retain information longer. For learning to last, it must move through all four connected systems (or stages) of memory, all the way to long-term storage. Reinforcing a memory at each stage helps it survive this journey intact.

How do you target each type of memory when designing an eLearning experience? Let's take a look.

Sensory strategies

The following high-level strategies can strengthen information that arrives in sensory memory. Remember these when writing storyboards, designing interfaces, or planning activities.

- **Consistency:** Consistent language and design help learners conserve energy. Use repeated terms when defining new concepts, create icons that always mean the same thing, and ensure interfaces behave as expected. Your learners shouldn't spend energy wondering why a link for "Broccoli Benefits" leads to a page titled "Fulfilling Fiber Fantasies."
- **But not too much consistency:** That said, it's vital to keep people on their toes and prevent them from tuning out. If you always use "Nice work" or "Try harder" as your eLearning feedback, users breeze by without paying attention—probably with an eye-roll for good measure.

- **Keep change meaningful**: Stay consistent, but not *too* consistent, right? It's a balance. But make sure the material still makes sense when you switch things up. You don't want to overwhelm or distract without reason. Some variation for the sake of attention is good. Too much random, meaningless chaos is DID YOU FORGET TO TURN OFF THE STOVE??? You get it. Choose your moments wisely.

Short-term and working strategies

Use these ideas to move information from the initial sensory stage into the short-term space. Short-term memories are more likely to be remembered if they trigger a change in a learner's prior knowledge, if they seem familiar, or if the information resonates as important.

- **Chunking**: Break information into smaller, digestible clusters to make storage easier. One example is the way you recall phone numbers. Instead of remembering nine separate digits, you probably chunk them into the area code (555), the city code (314), and the individual code (1592). Your brain holds this as three things instead of nine. Less effort, right?

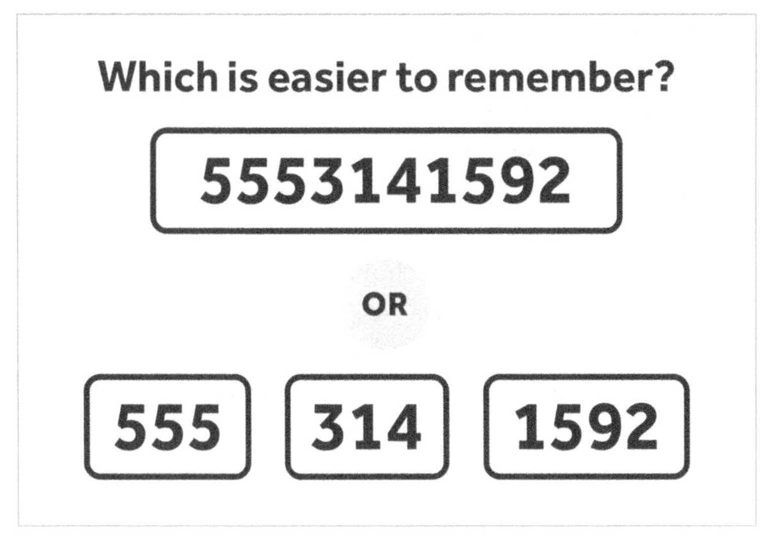

For extra stickiness, read the numbers aloud to rouse multiple senses.

- **Auditory stimulation:** Research shows that audio enhances short-term memory. This may be why music is such a powerful memorization tool and why advertising jingles will haunt you forever. Using voice-over and other auditory cues in your eLearning can reinforce information and move it to the next memory stage.
- **Use it or lose it:** Give learners a chance to apply their new knowledge so it doesn't fade. Games and puzzles help people exercise what they've learned. Active participation strengthens memories before they can dissolve.

Long-term strategies

Seal the deal with these methods so your learners store information for the long haul.

- **Repeat, repeat, repeat:** The more someone hears, sees, or does something, the stronger the neural connections become. Repeating objectives and summarizing key insights cements memories over time. That *Chapter takeaways* section I use in this book? It ain't just for funsies!
- **Everyone loves a story:** Humans are emotional beings, and they love a narrative—especially when it relates to their own lives. Not only does storytelling make visualization easier, but it can transform a dry concept into something more interesting. Relatable stories also connect the semantic (automatic) memory with the episodic (personal) for better reinforcement.
- **Scenarios and simulations:** Immerse learners in realistic situations to let them use what they've learned. Practicing in a safe environment helps solidify and deepen their knowledge.

Other practical tactics

We've talked about memory systems and ways to reinforce learning along its route. Let's wrap this up with a few tactics for designing your learning content. Piles of research support each method, so if you're curious about data and rationale, I encourage you to explore on your own.

- **Pose questions:** Ask questions rather than passively stating information. This encourages learners to consider their own answers before you supply one, and engages those mental muscles. Which working memory strategy does that remind you of?
- **Introduce spacing:** Learning is most effective when it's not a one-time event. It's best to expose people to information, give them a break, then prompt them to

recall and use it again. This approach is sometimes called *spaced repetition* or *distributed practice*.

- **Add healthy stress:** While you should apply this method with care, research shows that raising stress levels slightly can help make information stick. Ever been daydreaming in class and had a teacher call your name? I'll bet that helped you focus. Just don't catapult your learners into full panic mode, or the benefits dissipate.
- **Involve other people:** Humans are social creatures, so allowing them to interact with others makes learners more engaged. Not only does it build feelings of comfort and support, but it helps encourage active learning and participation.
- **Interleaving:** Alternate between different tasks or topics within a learning activity. This helps avoid monotony and keeps the brain on its... um... toes, so to speak.
- **Real-world examples:** How does your material apply to the world at large? Make information relatable by placing it in context.
- **Games:** Games are excellent for engaging learners and making lessons more entertaining. Use them to exercise new skills and practice retrieving information.
- **Quizzes:** Quizzes are par for the course (pun intended). They can boost self-esteem and reveal where your learners need additional instruction.
- **Feedback:** People want to know how they're progressing along the way. Offer constructive criticism and provide specific, actionable feedback.

Cognitive load theory

We've covered a lot in this chapter, so you may relate to this one. *Cognitive load theory* states that our brains can only process a certain amount of information at one time. When teaching something new, it's best to keep the cognitive load low to maximize understanding and retention.

Some of our earlier strategies help manage cognitive load effectively, including chunking information and simplifying complex concepts. It also helps to avoid simultaneous or unrelated tasks that might distract your audience.

People have limited attention for each session, so keep information focused and relevant. This impacts how you write scripts, design interactions, and plan the length and spacing of your training.

Cognitive load theory also reminds us to think about the limits of our audience. Teaching kids to boil penne, for example, should be vastly different from training forensic anthropologists... hopefully.

Exhausted? Good. Clear out that cognitive load. This chapter is done!

Chapter takeaways

Well, *almost* done. Just make sure this chapter sinks in before you continue. We'll be building on these concepts later.

- The brain employs several connected memory systems: *sensory*, *short-term*, *working*, and *long-term*.
- Instructional designers aim to reinforce knowledge at each stage, facilitating the learning process and

allowing memories to survive all the way to long-term storage.

- Various tactics have been shown to bolster memories in eLearning. Keep these methods handy as you begin project work, and feel free to mix them up.
- *Cognitive load theory* suggests that people can only process so much information at once. When designing a learning experience, reduce cognitive load and keep tasks manageable.

3

RECIPES FOR EFFECTIVE LEARNING

Now that we've discussed ways to help learning stick, let's go deeper by looking at two well-worn concepts in the instructional designer's tool kit: *Knowles' Principles of Adult Learning* and *Merrill's Principles of Instruction.*

Yeah, I know. The moment we assign official titles, everything becomes stodgy and academic. Just think of these as recipes for effective learning, along with recommended ingredients.

In the previous chapter, we looked at reinforcing memories at each stage along their journey to long-term storage. That's a convenient—but very neurological—approach to memory. But what if we took a more empathetic perspective?

As you read through these principles, look closely and you may recognize some of the memory tactics from earlier. If you can find common threads between learning methods, you'll build a deeper understanding of the ideas behind them. Then you'll be able to apply them in your workflow.

Knowles' Principles of Adult Learning

Malcolm Knowles was an American educator who offered several theories on adult learning. He summarized his views in the concept of *andragogy*—roughly translated from Greek roots as "leading men," as opposed to *pedagogy* or "leading children" (Knowles, 1984).

Knowles' Assumptions

There were two sides to Knowles' thinking. First, he suggested that adult learners share characteristics that are different from children. Namely, that adults...

- ...are autonomous and self-directed.
- ...come with life experience that can be considered in the learning process.
- ...look for rationale. They want to know why material is relevant and important to them.
- ...are problem-solvers, driven more by practical application than abstract theory.
- ...are self-motivated to learn and require less external prompting.

Knowles' Principles

He then used those assumptions to write practical learning principles. To create effective learning, adult learners...

1. ...should be allowed to participate in the direction and content of their learning materials.
2. ...learn better from experience and mistakes, rather than passive exposure to information.

3. ...are more engaged by material they consider relevant to their career or personal lives.
4. ...are confronted with problems to solve.

Applying Knowles' Principles to eLearning design

So how do these apply to eLearning design? Here are some thought-starters.

Give learners freedom

Giving adults autonomy in their learning can take many forms. In practice, think about creating options and enforcing fewer guardrails so learners can explore. You might ask learners to choose their starting point or let advanced users skip the basics.

In the case of games or branched scenarios, you could provide less up-front context, prompting learners to self-direct and uncover insights on their own.

Just be sure to balance freedom with support. Whether that takes the form of post-training peer discussions, summary insights at the end of a lesson, or the ability to ask questions for guidance, make sure your learners ultimately arrive at the right place.

Build on existing knowledge

As we've seen in earlier chapters, learners retain new information faster when it's linked to existing knowledge. You might remind learners of insights from a previous module before building on those ideas. Or you could use analogies in your material—introducing foreign concepts by comparing them to something familiar.

- Creating origami is like folding laundry.
- A freestyle stroke feels like reaching up for a light bulb, then putting it in your pocket.
- Surfing is a bit like skateboarding, except with more sharks!

One consideration here is that existing knowledge will vary by audience. If you're teaching electrical engineers about circuit design, their starting points and references will differ from the general public. This is a great reason to assess your learners' abilities before designing a curriculum.

Give them good reasons

Make it clear from the start why information is useful and why it's personally relevant to your audience. If learners don't understand why they are doing something, they'll be less likely to invest energy. Give clear examples of how people will benefit and what they can do with this knowledge.

Let them practice and collaborate

Allow learners to discuss and apply new material as a group. This fosters a team environment and enables them to practice their knowledge and skills.

Building a collaborative setting can be as simple as implementing a forum where learners can ask questions, or introducing team-based challenges.

Make scenarios realistic

Depending on your learning format, you'll want to write accurate, real-world scenarios so learners can practice safely before applying new skills.

Astronaut training springs to mind here. How would you deal with a failing life support system, malfunctioning CO_2 scrubbers, or fires in space? Best to find those answers in a simulator before flying off-planet!

If you've been paying attention, this can also scratch that problem solving itch and create the healthy level of stress from the previous chapter. Although admittedly, equating "space fire" with "healthy stress" might be a stretch.

Use productive failure

Finally, remember that mistakes are some of the best learning tools. Feedback and guidance are critical, especially when changing behaviors. If learners take a wrong turn, don't just leave them hanging. Tell them what went awry and how they can improve. Let your audience fail and see the consequences of their choices.

Merrill's Principles of Instruction

Much like Knowles, M. David Merrill was an educator who explored what made learning effective. He sifted instructional theories and, from common threads, distilled five principles of instruction.

In general, Merrill believed learning was most potent when it involved the audience in meaningful tasks and problem-solving (Merrill, 2002). This is consistent with everything we've learned so far: memories form when a person invests energy in the form of attention, then applies what they've learned. Merrill's notion of task-based instruction checks all the boxes and is particularly relevant to eLearning, where interactive tools are legion.

According to Merrill, learning is most effective when it has the following characteristics:

1. **Task-Centered:** Engage learners in tasks that require problem-solving and active participation. We've already discussed ways to introduce this in eLearning through games or branched scenarios. You'll also see this principle called *problem-centered.*
2. **Demonstration:** Show learners specific concepts or solutions. This is especially helpful for long explanations you might hasten with a real-life example. In eLearning, this could mean tutorials, step-by-step diagrams, infographics, or video walk-throughs of a scenario.
3. **Application:** Allow learners to apply what they've learned and reinforce mental connections. This helps them practice and find pitfalls they didn't consider earlier.

4. **Activation:** Relate new information to pre-existing knowledge. This makes connections faster and builds upon what learners know.
5. **Integration:** Help learners integrate new skills into their lives and interpret data for personal use. I can teach you to use the stove, but once you start flipping omelets, that's integration.

Common ground

Does a lot of this sound familiar?

- Invite learners to apply information rather than just throwing facts at them.
- Build upon pre-existing knowledge to speed up or layer the learning process.
- Create interactions that force learners to use new material in realistic scenarios or their own lives.
- Offer specific, actionable feedback along the way.

You'll see plenty of commonalities throughout this book. Not only because instructional designers build upon previous insights, but because there are trends in how humans learn.

Chapter takeaways

In this chapter, we looked at two tasty recipes for effective learning. We identified common themes and found tips to shape our material.

- To create impactful content, instructional designers use several methods based on memory and human learning.

- *Knowles' Principles of Adult Learning* list strategies for effective learning based on how adults learn best.
- M. David Merrill studied a range of instructional theories and compiled the recurring, choice bits in *Merrill's Principles of Instruction*.

4

MODELS AND FRAMEWORKS

Many arrive at instructional design from other professions—teachers, course creators, trainers—but the formal discipline has been around for a while. And while some techniques may sound Fancy, Academic, and Overly-Complicated (ooh la la!), they can all be broken down and used in the workplace. Remember, these aren't philosophical exercises; they're meant to help you build something.

This chapter will introduce a few standard models for planning your ID projects. Don't worry about memorizing details right now. Consider this your whirlwind tour rather than a cramming session. You (probably) won't be quizzed at the end.

Structuring your projects

Most ID workflows share a similar theme. Generally, you're going to:

- Figure out the problem or need.

- Define your project goals and approach.
- Design and build something to accomplish those goals.
- Put it into practice.
- Test the heck out of it to see what's working and what's not.
- Adjust accordingly.
- And if you're ambitious, you may even iterate throughout.

With that in mind, let's look at some of the frameworks you'll find along the way.

ADDIE

ADDIE is the most common and well-known instructional design model. It was developed in the 1970s at Florida State University for the U.S. military (Branson et al., 1975). Each letter of the acronym defines a phase for creating learning content.

- *Analyze*
- *Design*
- *Develop*
- *Implement*
- *Evaluate*

Models with this sequential approach are called *waterfall* because the steps move in a straight line, one flowing into the next.

In ADDIE, stages are intentionally broad to suit a range of activities. For example, in the Analyze phase, you might assess learner needs, conduct SME interviews, define project goals, and perform other tasks that shape your project. During

Evaluate, you may test how learners improve after training or run surveys for feedback.

Traditionally, ADDIE was a much more rigid and formalized system, but these days you might reference it as a loose project workflow. Designers often combine pieces from ADDIE with other models to suit their needs.

SAM

SAM is another design acronym. It stands for *Successive Approximation Model* and features a cyclical workflow that differs from ADDIE's linear approach. Developed by Dr. Michael Allen of Allen Interactions, this model is far more agile, allowing designers to make changes and updates quickly (Allen & Sites, 2012).

The key to this process lies in the word "approximation." Instead of expecting a polished result after one cycle, the idea is to build something for feedback, testing, and improvement before moving to the next cycle.

Unlike ADDIE, the letters here don't represent steps in the workflow. With SAM, you begin with a *Start* phase, then move to a cycle of *Analyze*, *Design*, and *Develop*. This sequence repeats until you reach an optimal solution, at which point you proceed to the *End*.

Rapid Prototyping

Well, this is awkward. Because in instructional design, *Rapid Prototyping* (with capital R and P) is sometimes touted as a formal alternative to ADDIE, with a few steps smushed together. In this workflow, users try out prototypes so designers can revise on the fly, identifying issues through constant testing.

However, rapid prototyping (lowercase) is just an activity you can do at any stage of the process. It helps flesh out ideas and gives learners a sense of content flow before building the live product. The purpose is to excite your stakeholders, try different approaches, and get early feedback so you can adapt.

Either way, prototyping is great stuff. We'll talk more about it later.

The Dick & Carey Model

What's that, you say? You like the structure of ADDIE and the cycles of SAM but want more detailed steps? Done!

The *Dick & Carey Model* is a popular approach to learning design. Like SAM, this approach—introduced by Walter Dick and Lou Carey (1978)—has an iterative flow that lets you cycle mid-stream until you arrive at the desired outcome.

This model outlines specific steps (although I'm combining a few for brevity):

- Setting instructional goals
- Assessing required and existing skills
- Defining learner objectives
- Developing tests
- Writing lesson plans and instructional materials
- Creating various evaluations

Unlike ADDIE, where evaluation happens at the end of a linear workflow, in Dick & Carey, you'll evaluate and revise constantly.

The design thinking approach

This one isn't just for instructional design. *The design thinking approach* is a framework for projects of all kinds. It's a problem-solving process that involves five high-level steps:

- Understanding the problem
- Brainstorming possible solutions
- Prototyping and testing
- Choosing the best solution
- Implementation

Hopefully, you're starting to see trends in all these models. And while this approach sounds pretty straightforward, it's most interesting when you think about skipping a step. For example, some teams try to save budget by omitting prototypes or delaying tests until everything's launched. Others ignore brainstorming and assume specs arrive fully formed with little or no research. Do you foresee any issues with those cuts?

So, while you can dig deeper, one takeaway is to ensure your scope includes time and budget for crucial phases. Gather research to inform your work, iterate to test your ideas, and choose your approach based on proven results rather than intuition.

How should you use all this?

The purpose here is to plan your project workflow at a high level before diving into nitty-gritty tactics. Traditional models won't prescribe storyboard templates or define your learning objectives, but they *will* prompt you to think about how to use prototyping, when to conduct evaluations, and whether your project lends itself to a linear or cyclical approach. They can help you find missing steps as you ease into planning. Practically speaking, they're also likely to pop up in interviews and client discussions.

There isn't a *best* when it comes to these methods. There's just the best for you and your project requirements. Go ahead and choose components of ADDIE, SAM, Rapid Prototyping, or other models you find, then adjust to taste.

Chapter takeaways

Models galore, right? There are plenty more where those came from, but you get the idea.

Knowing these industry standards should make them less daunting when they appear in the wild. They offer starting points and guides to customize your workflow.

In the next chapter, we'll prepare you to do just that. We'll look at tips for clients and schedules, choose tools you'll need, and raise questions you'll want to ask early on.

But first, a quick recap.

- Instructional design models help define stages of a learning design workflow. While you're sure to find complex diagrams and descriptions out there, remember they're all meant to build something.
- *ADDIE* is one of the most established ID models. Developed for the U.S. military in the 1970s, the acronym outlines steps for creating effective instruction.
- *SAM* is another well-known model that follows a cyclical process. It's useful for iterating quickly and adapting products based on testing.

- Many other ID frameworks evolved from people changing existing models to suit their needs.
- When considering the models at your disposal, step back and review your requirements. Grab what's useful, then fine-tune.

5

GET ORGANIZED AND CHOOSE YOUR TOOLS

When starting a project, you'll want to prepare a few things so nothing derails your momentum. Some notion of stakeholders, tools, and constraints is helpful, even though specifics vary wildly. Just remember we're building your baseline, so if you don't fully grasp it all right now, don't sweat it!

Also, the questions from this chapter are meant for your early project briefings. These might take place in client discussions or internal team meetings. Cherry-pick what's appropriate for your situation and get the answers you need. For your notes-free convenience, I've included them in the *Collected Questions* at the back of this book.

Figure out who you're working with

One of the first things you'll do is identify all the people contributing to your project. You'll decide how to interact with each group and what information you need from them.

Let's look at some of these roles and see where they fit in your workflow.

Subject matter experts (SMEs)

As the name implies, a *subject matter expert* is considered an expert in a particular field. They may be an academic researcher, a highly skilled professional in the industry, or a seasoned veteran with years of experience.

A subject matter expert can be anyone an organization or institution relies upon for sound, informed expertise. If you've been stranded on a tropical island for eight years, some might consider you a SME on cracking coconuts, even though you don't have a Ph.D. in the subject (although Doctor of Coconuts does sound pretty sweet).

SMEs are some of the most important people to speak with when designing your eLearning content. They offer insights into what others need to learn and serve as sounding boards when you draft material.

You may also enlist multiple SMEs with unique areas of expertise. One person may be an authority in a skill employees need to learn, while another has deep knowledge of the audience.

We'll cover working with SMEs in-depth, but in your first discussions, the task is to identify them and call out their specialties.

As a starting point, you might ask:

- Who knows the most about current issues and goals?
- Who can help define the business objectives?
- Who has expertise in existing material or relevant topics?

- Who can help me understand the target audience?

Whatever their specialty, SMEs can assist with content and provide valuable feedback throughout the design process.

Clients and primary decision-makers

A *client* is anyone who hires you or pays for your services. It might be a company, an individual, or a group. But as you work with clients, you need to distinguish your project's *primary decision-maker(s)*, not just the team members you met during kickoff meetings. This distinction is so critical that it's common to add "Assign primary decision makers" as a milestone.

Think that's excessive? Well, here's the issue.

Say you work closely with one client group during the early phases of a project. Their input directs your efforts and you revise deliverables based on their feedback. After investing time and effort, a primary decision maker pops in (often with little context or involvement) to overrule all the existing decisions. Now you're in a costly pickle.

Assigning a primary decision maker is a crucial step, and if you don't settle this early, it can have disastrous effects on your timeline and budget. Once you've identified your decision maker, be sure to invite them during reviews. Otherwise, you risk re-doing work.

If a client has an entire group responsible for feedback, you'll want to suggest a process that ensures timely, consolidated answers. A helpful exercise is to walk them through a scenario they'll see later. For example: "Assume I sent you round 1 of a storyboard. How will you deliver your consolidated feedback, who will provide it, and how many days after delivery can I expect it? If there are conflicts in the feedback, who will resolve them?"

When you're only working with one person, this is pretty straightforward. But if you have several voices of equal weight, these questions raise healthy debates you can help resolve.

Stakeholders

A *stakeholder* describes anyone with a guiding role who will contribute requirements or feedback. The primary decision-makers and subject matter experts are usually stakeholders, but this superset can also include investors, upper management, and even the learners themselves. Obviously, some of these groups have more clout than others, but you'll want to identify stakeholders up front so you can account for the effort you'll invest with them.

How much of your research will you need to convert into stakeholder-friendly materials? Should you reserve some time for convincing, persuading, or creating presentations for SMEs and higher-ups?

Be sure to plan for this time in your schedules and estimates.

Define your technical approach

Now that you know all the players, let's look at ways to build your eLearning and limitations to consider.

We'll summarize this with a few thought-starter questions to avoid getting sidetracked too early. And no, they don't all have to make sense to you yet!

How will you build your eLearning experience?

Technologically speaking, your eLearning experience could take many digital forms and use all kinds of media. Unfortunately, there's often a gap between what you'd like to build, what your clients can afford, what systems are already in play, and what you have the ability or staff to produce. With that in mind, it's helpful to outline your technical limitations before starting the design process.

The first unknowns are: who will develop the product you design, and what technology are they using to build it? These answers define boundaries for your interactivity and learning material.

Here are a few questions to help you and your clients talk through constraints.

- Will your eLearning involve custom development?
- Do you have access to someone (yourself, a team, or a specialized resource) who can build whatever you design?
- Are you or your client bound by an existing learning management system (LMS)?
- Is it part of your job to decide the best platform for your client, starting from scratch?
- Will you use eLearning authoring software (discussed shortly) to create lessons and interactive material?
- Are there features that might be challenging with your chosen technology?
- How will you track and analyze learners' performance?
- What digital devices or platforms will you support?
- What are your accessibility requirements?
- Should you plan for multiple languages or regional content?

These bullets mark the tip of the requirements-gathering iceberg, but hopefully, they get your gears turning. Some projects even have a dedicated technical discovery phase where you can ask questions like these.

It's okay if these concepts sound alien right now. By the end of this book, they'll be old hat. The gist is that if you want to make achievable designs, you need to understand the strengths and limits of your tech.

Are you using a learning management system?

A *learning management system* (LMS) is an online platform designed to support learning and instructional activities.

Typically, an LMS offers tools that allow instructors to create learning materials, assess student progress, collaborate with colleagues, and track student performance over time. Details vary from one system to another, but going in this direction can be a major investment.

Maybe your client is already committed to one, or perhaps it's your job to help them choose a good fit. But once your client is bound to a specific LMS, you'll probably design your experience to work within that environment.

Again, that means you're not designing a curriculum from a blue-sky approach, because you need to review what's possible within the boundaries of the LMS toolset. For example: if your system only supports video playlists, you don't want to design a curriculum based on drag-and-drop matching games.

Most platforms include some combination of course management, discussion forums, assessments and analytics, surveys, and multimedia.

The features of each LMS can be elaborate. As with any system affecting your work, you'll want to educate yourself before designing.

How will you create or acquire multimedia?

We'll look at multimedia in the storyboarding chapter, but some questions to consider now are:

- What types of visual assets will you use in your learning experience?
- How will you create or acquire video, animation, photography, illustrations, or iconography? Are you producing them yourself or paying for stock media?
- Does your eLearning authoring tool (if you're using one) offer character or illustration libraries?
- Does your client have existing assets they want you to use, or will everything need to be custom?
- Do they have the budget to purchase or create whatever assets are required?

If you don't know these answers already, it's a good conversation to have with clients before you agree to a scope of work. Any stock assets you need to buy, and the time you need to source them, should be included in the budget.

Choose your digital tools

Okay. This is a tricky one.

As I mentioned at the outset of this book, recommending specific apps is a risky business that instantly shortens the lifespan of this material. One day you're providing helpful references; the next, your advice is dated, irrelevant, and, in the worst case, misleading.

So, rather than focus on software that will change, be updated or acquired, lose favor, or evaporate altogether, I'll outline tasks we need software to accomplish and traits to look for in an offering. This should help you explore digital tools and make your initial picks, even as technology changes.

That said, feel free to peruse the *Tools & Resources* section at the back for some starting points.

Mind-mapping and action mapping

The purpose of any mapping practice is to capture thoughts and visualize relationships between them. When evaluating software for this purpose, look for a few crucial features.

First and foremost, software should allow you to draw connections between ideas and build diagrams showing links or progressions. It should let you organize in various ways, either by grouping related nodes or prioritizing elements based on hierarchy. You'll want to show sibling and parent-child relationships. For example, if your main topic is animals, you might add vertebrates and invertebrates as two child nodes. You could further break down vertebrates into birds, mammals, fish, etc.

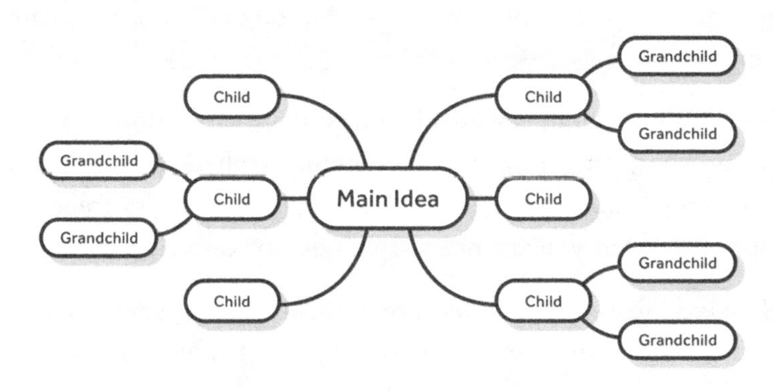

Second, mapping software should be intuitive, with controls that don't inhibit your speed. Remember how quickly the brain forgets tenuous information? You should be able to snatch ideas before they dissolve, then reorganize or prune them later.

I used MindMeister for examples in this book, but you'll find options in *Tools & Resources*.

Potential uses:

- Action mapping (discussed later)
- Capturing and grouping brainstorm ideas
- Organizing topics into sequences or chapters
- Grouping user profiles or learners
- Creating site maps
- Visualizing branched scenarios

Sketching

In this context, we're talking about *sketching* as any method that lets you quickly capture free-form visual ideas. That could be pencil and paper, tablet and stylus, laptop and mouse, and any combination of software.

You might gravitate towards features like style, options for nibs or brushes, and drawing functionality—whatever suits your preferences and abilities. Or you may prefer a less freehand option that lets you arrange shapes and text on a canvas.

The only feature you *must* have is the ability to sketch quickly in a medium you can share with others. Everything else is a bonus.

Sketching is handy for instructional design because it allows you to visualize ideas quickly and experiment with different approaches. You might reach for your stylus when doing rough wireframes, laying out your UI, doodling a user flow, or scribbling napkin sketches for your storyboards. By restricting yourself to raw lines and shapes, sketching removes the temptation to get sidetracked by visual polish at the wrong stage. It helps you focus on problem-solving rather than presentation.

Sketching is also convenient for brainstorming and saves time explaining what's in your head. If you spend more than a minute describing a visual solution, just draw a picture and save yourself a thousand words.

And while digital tools make sharing ideas or revising work easy, don't underestimate the value of a hand-drawn sketch on paper or whiteboard—there's no learning curve, and you probably have all the supplies you need.

Potential uses

- Rough wireframes or UI layouts
- Storyboard thumbnail sketches
- Working sessions to visualize ideas and tag-team solutions

Visual and UI design

The tools we're looking for here may appear under visual design, graphic design, or, more specific to eLearning, UI design. As you'd expect, instructional designers use these programs to create visual assets and artifacts like layouts, interfaces, style guides, and even wireframes.

A word of encouragement and caution: this is one of the most ephemeral software categories for designers. So, I'm going to offer two conflicting pieces of advice:

- Invest the time to master your software of choice; then you can focus on refining your work rather than deciphering tools.

- Industry standards are fickle, so don't get attached to a particular application. Keep your eyes peeled for new tools and adapt, adapt, adapt.

With so many software options for UI design, finding a good fit can take time. You may even want several design apps in your arsenal for different tasks.

Vector-based software is great for interfaces and wireframes, while raster-based applications optimize and manipulate imagery. Some contenders have online collaboration tools, which make team sharing and client feedback easier.

Look for features that suit your requirements, subscription fees that match your budget, and any conventions from your workplace. Fortunately, most products offer free trials or tiers before you commit to purchase.

While this book's *Tools & Resources* lists a few options, an online search for "UI design tools" will also get you rolling.

Potential uses:

- Layout and interface design
- Style guides or UI kits
- Optimizing images
- Exporting visual assets for development

Documentation

This one requires little explanation. Just pick your software for long-form, professional writing. Preferably, your choice will allow easy sharing (with some level of security) and permit others to comment.

Personally, I love apps that let me write in markdown and drag blocks without copying and pasting, but choose a tool that feels right to you. You'll also want something that allows you to work in tables and sheets; it can be part of the same app or a close sibling.

Scripts, requirement docs, outlines—you'll use this as expected. Be sure to set clear file and folder conventions, so you and your team can find things later.

Storyboarding

Storyboarding is an essential phase of the ID process that results in a blueprint for your narrative. You might develop your storyboard as a document, a slide deck, or even sketches. The important thing is that it outlines the path of your eLearning experience and the supporting content.

The software you choose will vary by the type of storyboards you produce. Some are more illustrated and include frames and thumbnail sketches.

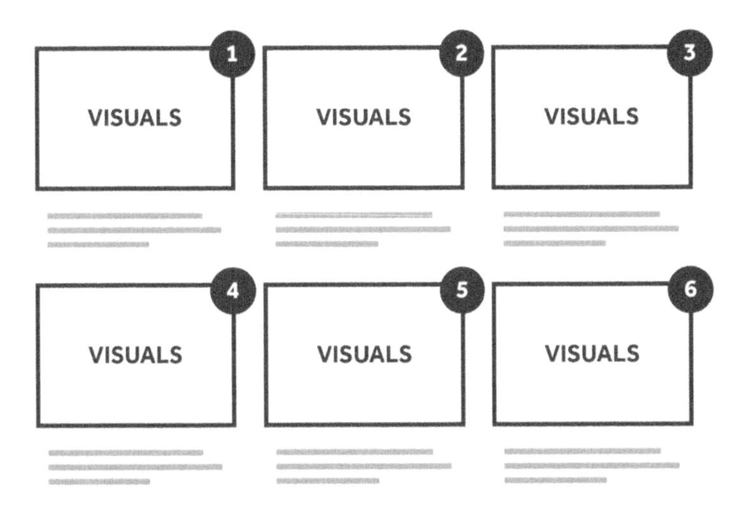

Others (like the one in this book) use a tabular format with cells for voice-over, images, and descriptive text.

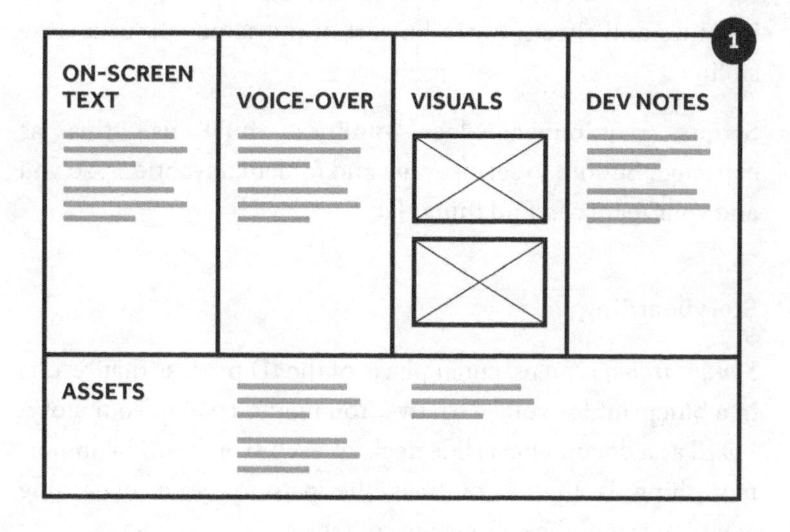

It's worth considering that for storyboards and other client deliverables, cloud-based software can make it easier to collaborate and send feedback, rather than screen sharing or passing files back and forth.

We'll cover storyboarding in detail. For now, if you've got something that allows you to use tables with text and images, you're all set.

eLearning authoring tools

I hinted at this in the technical approach section, but it's a biggie.

In some cases, eLearning is designed and developed almost entirely within *eLearning authoring tools*. Instructional designers use them to create content, assemble lessons, and add animation or interactivity. Generally, eLearning is built within an authoring tool, then uploaded elsewhere—possibly an LMS or dedicated server. With the exception of visual design and hosting, these tools can serve as a one-stop shop for eLearning projects.

At a minimum, they usually have a canvas to lay out design and media, a way to navigate between slides, and UI to build links and control a learner's path through the experience. The workflows may seem loosely familiar if you use presentation apps.

Some offer illustration and character libraries, pre-designed slide templates, or tools for video recording and screen capture. The features you need depend on the format of your eLearning. Are you creating a series of talking-head videos or a mixed, interactive course?

Keep in mind that not all projects leverage authoring tools. If you're developing an experience from scratch or using a standalone course-building platform, you might not use them at all.

When choosing software, I offer a few tips.

- Get a sense of the learning curve and ease of use.
- Ensure the features meet your requirements for animation, media, interactivity, VR environments, etc.
- Evaluate image, character, template, and interface libraries.

- Make sure software is available for your operating system.
- Look for responsive design and accessibility support.
- Dig into deployment and hosting options—in other words, how does your eLearning experience move from the authoring tool to its destination online or elsewhere?
- Take a peek at the pricing plans. Costs can vary widely.

Again, visit *Tools & Resources* in the back for inspiration, but remember, these tools will change rapidly.

Chapter takeaways

I realize this is a lot to mull over before a project's even begun!

Much of this may sound familiar if you're a seasoned instructional designer. But if you're new, just breathe and relax. Don't expect to have everything figured out on day one. In fact, if you have very few of these answers right now, that's perfectly fine. The idea is to discover who's involved, define the technology in play, and make some early choices about your tools.

Here are a few key points as you're prepping:

- Identify the people who will influence your project and provide feedback. This includes *subject matter experts (SMEs)*, *primary decision-makers*, and general *stakeholders*.
- Define your technical constraints and considerations —from the eLearning experience you want to deliver to your multimedia assets.
- If you're working with a *learning management system (LMS)*, make sure you understand the platform, its benefits, and its limitations.

- Preselecting digital tools can save time when you're working. Look for software that supports mapping activities, sketching, visual design, documentation, and storyboarding.

6

A TASTE OF PROJECT MANAGEMENT

Since instructional designers often direct other groups, knowing project management basics can be handy. You'll usually fill in the same blanks: what do you need to do, who is involved, and how long will each step take? Whatever your situation, we'll check out some standard practices.

If you're absolutely *itching* to dive into design methods, feel free to skip and revisit this chapter later. Schedules and resource plans are critical at a project's outset, but they're also easier once you've learned the stem-to-stern roadmap.

So, read on without stress or return here at will. Either way, you know where to find this when you need it!

Pick your project management workflow

There are plenty of ways to structure a project, and if you work at a company, your employer may already have an established workflow. But if you're operating independently, there are two popular options. Both offer benefits and drawbacks, making them suited for different projects.

Waterfall

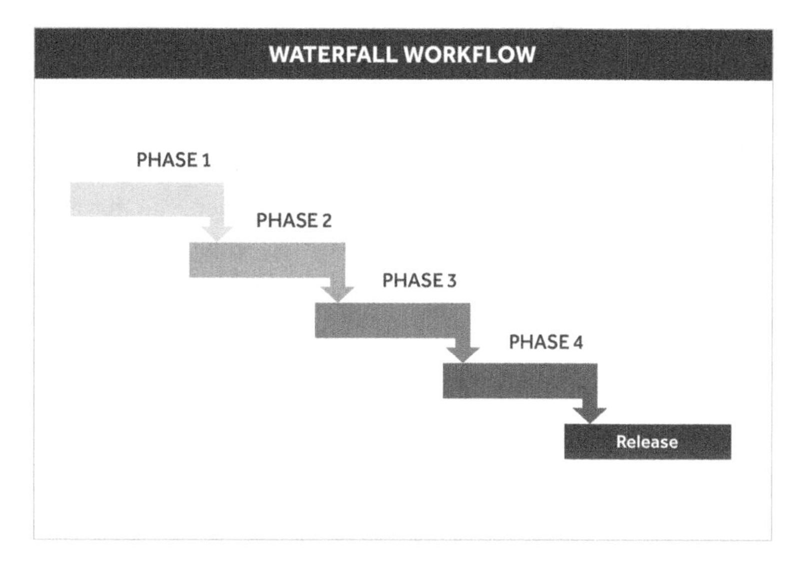

Let's start by looking at the *waterfall* workflow. This method takes a traditional approach, addressing each stage in linear sequence. With this method, you determine the goals and timeline up front, then move through each *phase* step by step until completion.

In practice, this works like passing the baton in a relay race. Instructional designers finish their storyboards and hand them to eLearning developers. Developers use these artifacts to build a functioning experience and deliver it for testing. And so on until launch. In theory, there's no cycling or revisiting prior phases.

This makes waterfall a good choice for projects with strict timelines and known requirements. You begin with a schedule, budget, and initial specs to develop a finished product. You'll mark progress with milestones and deliverables.

However, changing course in a waterfall project can be like trying to whip U-turns in a loaded school bus. Since each phase builds upon its predecessor, making sharp pivots means revisiting all the plans you made in advance. This can trigger renegotiation, change orders, and other slow-moving conversations before work can proceed.

If this flow reminds you of the ADDIE model, you have an excellent memory! However, in this case, you'll include any project phase you want in the plan, not just stages from the ADDIE acronym.

In general, waterfall workflows are great for fixed budgets and stable due dates. But unless you accommodate time for experimentation in the schedule, you lose some ability to adapt along the way.

Agile

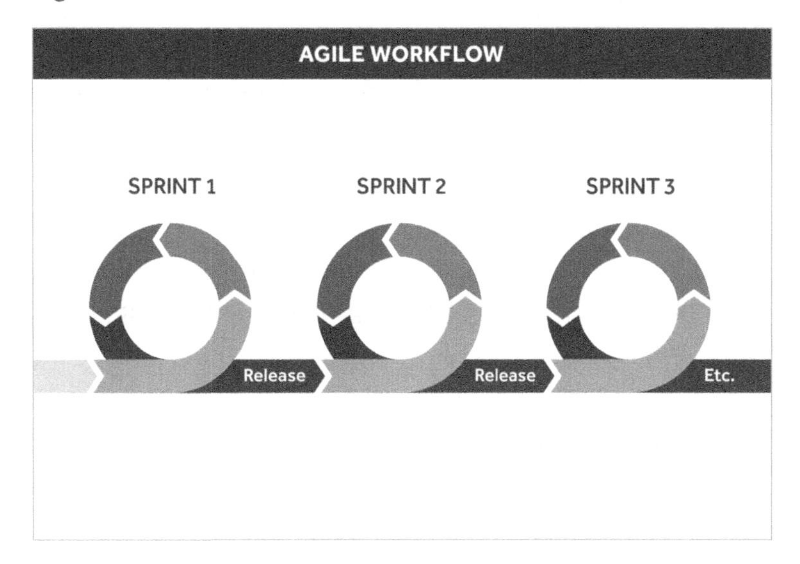

On the other hand, an *agile* workflow divides projects into smaller cycles or *sprints* that you can tackle iteratively. Each sprint includes objectives and reviews, and based on how things go, the team may refactor subsequent sprints.

Did you get user feedback that convinced you to pivot or reprioritize? Did you finish only half the planned tasks and need to revise work for the next sprint?

Agile allows you to respond to change by reassessing each cycle. It works especially well for projects that involve rapid iteration, like software development. It's also useful when scope or requirements are in flux or when you need to weather changing conditions. However, this flexibility comes at a cost.

Since sprints are constantly revised, it may be tricky to lock down how long you'll need to finish a project. In fact, the definition of finished may even change over time, making it difficult to define an end date—or a budget—in advance.

On the scheduling front, it's also hard to predict when resources will free up for other projects. This is another reason an agile approach is common for product teams, where members can focus on one thing rather than hopping between clients.

For the sticklers, I'll note that regimented Agile workflows (note the capital A) exist for software development. These methodologies have their own systems and specialists. Here, however, we're just talking about a general, lowercase approach for iterating within a project.

Which workflow should you use?

If you have a fixed deadline and budget, a waterfall workflow may help you reach a pre-defined, predictable outcome. If your priority is to test and refine often—with some wiggle room in those areas—the flexibility of an agile workflow could serve you well.

In reality, many companies adjust these workflows to suit their needs. A strict agile methodology is often expensive because, in theory, the final budget, timeline, and result can be unpredictable.

Similarly, projects are rarely 100% waterfall because some change is expected. Often, companies using this method still accommodate pivots—they may just charge extra for any overages.

One helpful trick is to walk through your project step-by-step, then imagine the impact when things don't go according to plan.

For example, what happens if your client misses a feedback deadline or asks for extra revisions? You could add another sprint in an agile workflow, but it may cost more and push the release. Do your clients have that flexibility, or is a timely, cost-effective launch their priority? What happens if your client runs out of funding before the project goes live?

Both workflows have pros and cons, so evaluate your needs and decide which best fits your situation.

Plan your schedules and resources

Project scheduling has a huge impact on the quality of your eLearning and the health of your team. Estimating how long things take and where they'll fit on a calendar lies at the heart of any workflow, but there's an art to the science.

Here are a few tips to help you plan.

Scoping and estimation

Developing a clear scope at the beginning of an eLearning project is crucial. You'll compile a list of all required tasks, estimate how long each will take, and assign appropriate roles. The goal is to develop a full view of the work so you can assess how long you'll need to finish.

An accurate scope helps you translate tasks into timelines and find discrepancies between budgets and requirements. It's also a good time to call out assumptions—what work is and is not included in your scope?

During this exercise, you should set expectations for team members and break tasks into chunks so they're easier to track. "Create wireframes for main navigation" and "Develop storyboard for module 1" are tasks that can be scoped and assigned.

Number of rounds

When you need a client to review something, the process follows a basic pattern. You work on a deliverable, share it with your stakeholders, and ask them to approve it or provide feedback for another *round*. Often, a client team needs time to assess work internally or present it to others.

This cycle can grow costly, and a common rookie mistake is to plan your schedule (and budget) as if everything flies through in one pass. So, before that happens, you'll want to clarify the following.

- How many rounds of work will your client want for each deliverable?
- What is their internal review process, and how many people does it involve?
- How long will your client need for each internal review?

Generally, planning for two to three rounds provides a healthy buffer for deliverables like storyboards. Once you have a starting point, you can chat with your client to see what they want to adjust. With your estimate in hand, it becomes obvious when they're trading revisions for a lower budget.

I'm not calling these "rounds of revision" because I've seen that phrase lead to major scheduling snafus. If you present a storyboard three times, how many rounds of *revision* have you gone through? That's right, two. The first presentation was not a revision.

In practice, the number of rounds tends to diminish as a project continues. Design conventions solidify, and you start reusing components. At some point, you're no longer developing a visual style or agreeing on the basics—you've locked all that down, and you're building on established rules.

Oh! And be sure to label each deliverable with the round number. When clients realize they're running low on time, their feedback gets much more specific.

Sprint or phase planning

Once you have a list of tasks and how long they'll take, you can start arranging them on a timeline. Whether you're going with waterfall, agile, or something in between, your schedule should align with your scope.

On waterfall projects, you'll group work into phases based on milestones, deliverables, or specific activities that make sense for your team. Agile projects, on the other hand, should iterate using sprints. And while sprints may change along the way, you can still set initial objectives.

This calendar planning gives you and your clients an idea of how long each stage will take and when to expect deliverables. You'll also establish a progression for your project and call out dependencies. Just as you wouldn't put your cart before the horse, you wouldn't put user testing before prototyping. What would users be testing?

Since this isn't a book about project management, we won't delve into project plans or Gantt charts. But as an instructional designer responsible for guiding others, it's worth considering how you spread work over a timeline.

You can't do everything simultaneously. So what does your schedule look like?

Set up key meetings and milestones

Creating a schedule with milestones keeps your client interactions predictable. You know your deadlines, your stakeholders know when to expect things, and everybody gets the details they need—simple, foreseeable, and anxiety-free.

Milestones represent meaningful points on your timeline. They might coincide with a deliverable, an internal or external review, or the end of a project phase. What's important is flagging them in advance and ensuring everyone knows their part.

You might set a review for storyboard approvals before recording narration. Or maybe you'd want a milestone to mark the end of design and the beginning of technical development.

Some milestones even happen at regular intervals. You'll undoubtedly have a kickoff meeting and design reviews, but you may also want client stand-ups every few weeks to adjust tasks for each sprint. Minor course corrections along the way are preferable to disaster at the end.

Three questions that may help with milestone placement are:

- Where might stakeholder feedback have the biggest impact?
- Are there long work blocks where a misstep could force you to backtrack?
- When will you need all your stakeholders to gather and agree on something?

Your early milestones don't have to be perfect. Call out obvious touch points and book participants' time so they expect what's coming. If your first guess has too many meetings or too few, you can always change it later. Just be sure to have these discussions early.

Resource planning

Among other things, the term *resources* in project-management speak often means staff members. So when you're scheduling resources or doing *resource planning*, you're figuring out who will work on what and when that will happen. Since resources are always limited, this helps schedule enough people for a given period, with the right roles and skills for all their projects.

The shape and size of this group may vary. You might be working alone on a freelance project and doing everything from research to technical development, or you may collaborate with a team that divides the work. In the latter case, it's common for specialists to engage in some stages more than others, and it's helpful to forecast this in your planning.

Either way, you'll want to avoid scheduling people before they have what they need. For example, suppose you'll be working with an animator. You probably don't want them to start before storyboards are approved because they'll have nothing to animate. You might even burn through all their time before they've logged a single keyframe!

The goal is to take stock of everyone you'll be working with, match available talent to your timelines, and involve the right people when it makes sense. Early planning helps you reach the finale without exhausting your budget or losing team members when you need them most.

Chapter takeaways

Hopefully, this taste of project management was more sweet than sour!

While planning requires some fortitude, the goal isn't to hinder you or kill your creative drive. It's to remove obstacles early and clear your path to the finish line. Instructional designers often pitch in by defining project scopes, schedules, and resources. Understanding the big picture can help.

- *Agile* and *waterfall* are two of the most common project management workflows, each with pros and cons. Waterfall projects are usually divided into phases and move linearly, while agile workflows build on iterative sprints.
- Plan your project by listing tasks, defining your scope, and plotting your work against a timeline.
- Agree on the number of rounds for each client-facing deliverable and use this information in your schedules. Label each document with a round number as you review work with stakeholders.
- Schedule meetings and milestones to keep your client interactions predictable.
- *Resource planning* ensures you have the right people assigned throughout your project.

7

GUIDE YOUR WORK WITH THREE QUESTIONS

Defining your audience and goals is crucial for any eLearning project, regardless of topic.

There are a few ways to think about objectives. You might consider what clients want for their organization or what learners need to improve. Sometimes these two things are in lock-step. Sometimes, they're wildly different.

And before you can do any of this, you need to know as much as possible about your audience. What skills or knowledge are they currently lacking? What training will they respond to, and what settings or environments influence their work? You'll use these insights to create tasks that move learners from where they are to where they should be and decide how you'll measure success.

We'll look at methods to guide this process in the following chapters. But first...

Who are your learners?

In instructional design, we need to understand our learners—their educational backgrounds, communication preferences, and relevant challenges.

These details inform every aspect of a project, but one of the most important is credibility. For training to be relatable, learners should feel you've designed content with them in mind. That could mean using specific language—tone, style, and word choice—or choosing exercises from their field. You'd use vastly different activities when training realtors than you would for AI software developers.

You can gather information about learners from various sources, including surveys, interviews, focus groups, and direct observations. Stakeholder interviews (especially those with subject matter experts) provide a wealth of knowledge.

What are the business goals?

Simply put, *business goals* are the overarching objectives that an organization sets for itself. These might include growing revenue or improving service quality. But they can also be more focused, like launching a product line or making more deliveries on time.

Some goals may benefit from training, and some may not. But here's the catch: there's often a difference between what your client says they must do and what serves their needs best.

What's the plan for learners?

Once you know an organization's training-related goals, the next questions are:

- What do learners need to learn?
- What should they be able to do?
- What will get them there?

We'll use a few methods to figure this out, and there are two terms you should know.

Learning objectives are goals we want people to achieve after completing an activity or module. *Actions* are steps learners must take to complete a task or learn a particular concept. We'll look at ways to define your learning objectives and actions. In practice, you'll likely be working with one or the other.

Sounds good. What's next?

Now that we have our key questions, let's talk about finding answers! In the next chapter, we'll interview our most valuable data sources: the SMEs and stakeholders.

8

WORK WITH SMES AND STAKEHOLDERS

Interviews with clients and SMEs form the bedrock of effective eLearning. But you can't just show up, grab a notepad, and start work based on what they tell you. Stakeholders are priceless, but they're not infallible.

The problem with interviews

We've already hinted at one challenge—your clients may not know exactly what they need. They could arrive with opinions or goals fully formed, or they may leave it all in your hands. Either way, you'll need to distinguish between what clients say they want and what will actually address their issues.

Subject matter experts leave you with a different problem: information excess. They tend to share heaps of data they consider essential. SMEs are, by definition, experts in their field, but they may be new to eLearning. While you want their insights, you may not be able to include everything they'd like.

We'll look at ways to guide discussions with SMEs and stakeholders, weigh their input against your findings, and distill the most potent content for your audience.

Let's take a deeper dive.

Lead with healthy skepticism

Your clients may arrive with an assumption that training is the solution to their problems. This isn't always the case.

Say a client walks in, hands you a stack of material, and asks you to build an online course. Well, what if their content has little to do with their issues or doesn't need to be memorized? What if they're introducing tools their learners would instantly understand? In some cases, you may have more impact by updating software, boosting morale, or posting quick-reference checklists.

Similarly, eLearning doesn't always apply (and yes, I hear your inner screams about this book title). If a problem won't improve through digital practice, for example, you may want to look elsewhere.

I mention this so you'll keep an open mind as you progress. Training isn't always necessary, and eLearning isn't the only format. Don't let a client's assumptions drive your approach.

Draft your business goal

Your clients are dissatisfied with something. Otherwise, they wouldn't be looking for help. One of your first tasks with SMEs and stakeholders is to identify the issue and write a clear business (or project) goal. Everything else stems from that.

Say your client owns a cupcake chain and arrives with the somewhat hazy request of: "We want to increase customer satisfaction."

Great. That gives you a starting point. It tells you customers are unhappy (or, at least, less happy than your client would like), but it doesn't reveal the underlying causes or who can resolve them. Clients don't always have these views well-defined, so you may need to help them look closer.

You might ask how they measure satisfaction. Do they use online polls? Ratings from external review sites? In-store comment boxes? Without a means of tracking, you can't tell where you started or how much you've improved.

In *Map It: The hands-on guide to strategic training design*, Cathy Moore offers guidance for writing a business goal (Moore, 2017).

First, you'll work with stakeholders to define a problem and find ways to address it. Then, you'll choose a metric for success, add a due date, and decide who's doing what.

Suppose your cupcake clients gather feedback through online surveys. They know customers are irked most by long lines and delays, and decide managers are in the best position to make changes. In that case, your goal may be:

Increase online survey ratings by 25% in six months after managers reduce customer wait times.

In this statement, you've described:

- A way to measure success using existing metrics and a target number.
- A timeline.
- The people who must act or adapt.
- Some notion of what they'll do.

If you're unsure about the last two bullets, that's okay. You may not know what people need to do at the outset. Your goal can evolve as you learn more.

And don't worry, we'll practice this method throughout the ID workflow.

Nine tips for solid SME interviews

Sitting down (in-person or virtually) with a SME is a great chance to get expert-level knowledge on a topic. However, you'll want to structure conversations to avoid raw, unfiltered information.

Below are some tips to get the most from your SME time. And as usual, questions from this chapter are stowed in the handy-dandy *Collected Questions* at the back!

1. Give them some context

If someone asked about your life, where would you start? You might ramble off a long chronology, beginning at birth and ending with you in the chair before them. Comprehensive, sure. But meandering. You'd have no idea what they're looking for in that jumble of data.

If, instead, they said they were writing three paragraphs for your company bio and asked targeted questions, you'd (hopefully) sift through your answers with more purpose.

The same principle applies to your SME interviews. Tell them what you're building and what insights are useful up-front. For context, you might:

- Walk them through your design workflow and development process.

- Discuss the eLearning technology you'll be using as well as its strengths and limitations.
- Show them examples of similar learning experiences to get a sense of the final product.
- Highlight the purpose of prototypes and artifacts you'll need them to review.
- Emphasize how much information learners can reasonably absorb within training constraints.

Giving SMEs context accomplishes a few things. First, it builds trust and makes them feel involved. Second, it directs their mental foraging to harvest the juiciest, most relevant bits of knowledge. And finally, it prepares them for what to expect as work begins.

You'll need SME feedback throughout the project, and not everything will make the final cut. Your work will go more smoothly if they understand why.

2. Distinguish impactful vs. extraneous material

Of all your SMEs' material, what should you include?

This may sound counterintuitive, but avoid asking for information SMEs want learners to know. That approach invites loads of content that may have no tangible impact.

Instead, ask SMEs about their hopes for the project and behaviors they want to change. What performance problems exist, what are learners doing now, and what information could help?

When reviewing material, have SMEs walk through specific scenarios.

- How and when would a learner use this in practice?
- What would go wrong if learners didn't know this?

- If every learner had this information, how would it impact the business goal?

If material impacts these areas, it may be worth including. Otherwise, you've justified a possible cut. Education may be its own reward, but unless you're working in academia, your project should have a measurable goal.

Also, ask for any previous training to see what's worked and what hasn't. This helps you avoid past mistakes and review their existing tech approach.

3. Structure your sessions

Create agendas, write questionnaires, and prepare surveys in advance to keep your sessions productive and professional. This removes scrambling and time-boxes each exercise, and you'll be more likely to get the answers you need.

Most importantly, you'll feel confident leading interviews if you've already plotted your steps. You've got a plan, and the SME knows they're in good hands.

Not that you want to stifle free-form conversation entirely. SMEs should have leeway to go off-script, and exploration can be valuable. But guidelines help focus your time and reduce anxiety.

Check out *Collected Questions* for some SME-friendly thought starters.

4. Fish for scenarios

Throughout your eLearning, you'll put learners in various situations to see how they react. Having a library of scenarios is useful for what-would-you-do exercises, quizzes, games, or branched storylines. Obviously, it's best to make these interactions as realistic as possible. And that's where your SME's experience comes in handy.

Ask about common issues, complaints they've received, and anecdotes that might resonate, either from the employee's or the customer's perspective.

- What problems have you seen, and what were the circumstances?
- What happened as a result of mistakes?
- Were there consequences or attempted remedies?
- What worked, and what didn't?
- What made situations stressful?
- Were there conditions, settings, or details that set the stage?
- What would successful outcomes look like in each case, and what actions could the learner take?

The more accurate your scenarios, the more your content mirrors reality. Not only does this build credible storylines, but it also prepares learners for what they'll encounter.

5. Ask about possible challenges

Ask SMEs about obstacles that could derail your eLearning. This might include problems they've seen in past activities, a lack of training time, spotty leadership support, or your learners' discomfort with technology.

This discussion can be open-ended; you want a sense of things to avoid, test, or plan for in your methods.

6. Schedule your SME time

Booking sessions early sets expectations and shows your SME that you value their time. Discuss the best days to collaborate, walk them through calendar milestones, and schedule check-ins or status updates.

You'll also want to decide when to work synchronously vs. asynchronously. Real-time sessions help minimize write-ups and misinterpreted feedback, but coordinating too often can be draining.

Generally, collaboration should be synchronous when brainstorming, presenting, or reviewing new deliverables. It can be asynchronous for feedback on familiar material. For example, you might work in real time when devising practice scenarios. Sharing notes on third-round storyboards, however, may not merit a meeting.

Whatever your cadence, scheduling SMEs in advance ensures their involvement. Just make sure your activities fit in the allotted time. Nobody likes a chronic under-booker.

7. Ask to record your sessions

You'll save a ton of effort by asking your SMEs if you can record their interviews. This lets you focus on the conversation rather than note-taking, and captures every crumb of information. It's better to spend your energy on active listening and off-the-cuff questions. You can always transcribe and annotate later.

Remember cognitive load, friend. Focus on one thing at a time!

8. Get consensus for major decisions

When client decisions have major consequences, make sure all your stakeholders are aligned. Milestones for goals, learning objectives, and storyboards are all possible pain points, and you don't want to rely on incomplete or conflicting feedback. If that means gathering the group to discuss, quibble, and align on a course of action, then go for it!

And yes, this one relates more to primary decision-makers than SMEs. But it'll save you some grief.

9. Don't be afraid to push back

Clients hire you to be the instructional expert. They expect you to challenge them when necessary, and (most of them) aren't looking for blind agreement. If a request doesn't suit the project goals, say so. Most clients appreciate feedback, even if you question their assumptions and offer alternatives.

Building on these basics

Now that you've got some ground rules for working with SMEs and stakeholders, we'll add assessments and exercises to help structure your working sessions. In the next few chapters, we'll dissect what learners need to do, map quantifiable goals, and apply everything in design.

I hope you're ready, and hold on to your SMEs!

Chapter takeaways

SMEs and stakeholders are the ultimate sources of knowledge for your eLearning experience, but they come with some caveats. In this chapter, we looked at tips for setting expectations, establishing your working cadence, and making the best possible use of your discussion time.

- Interviews are gold mines for content, but watch out for pitfalls. As an instructional designer, you must distinguish between what stakeholders tell you and what's relevant for learners.
- Begin your projects with healthy skepticism, and use research to form your approach. Don't automatically assume training is the best solution for a client's needs —even if your client does.
- Work with clients to write a clear, quantifiable business goal to help direct your efforts and measure success.
- Give your SMEs some context up-front, so they know what you're building and the kinds of information you're looking for.
- When evaluating material from your SMEs, help them filter content that will impact their goals. You'll also want to review existing training for insights.
- Structure your SME sessions by preparing questions, materials, and activities in advance.
- Your SMEs are excellent scenario sources. Pick their brains for real-world examples you can use for training.
- Poke around for known issues that might impede the learning process. If your SMEs have obstacles or caveats to share, now's the time!
- Book your SME meetings and ask to record their sessions.
- During key milestones, make sure all primary decision-makers reach a consensus before you move on.

9

ANALYZE YOUR LEARNERS, NEEDS, AND TASKS

Ultimately, your job as an instructional designer is one of movement: to help learners get from where they are now to where you want them to go. Once you've identified the gaps, you can design solutions to fill them.

Enter the *needs assessment* and *task analysis*.

A *needs assessment* identifies what learners need to know, have, or do to address a performance issue. Here, you'll examine their existing knowledge, instructional preferences, skill levels, and motivation. You'll look at the delta between their current situation and the objectives you set for them.

Once you've defined what needs doing, a *task analysis* breaks complex tasks into smaller, more manageable steps. This lets you verify instructions and build activities that suit your audience.

Recalling our earlier description of the ADDIE model, these two exercises fall squarely in the Analysis phase and are often used in tandem.

Conducting a needs assessment

In the previous chapter, you helped your client write a business goal—outlining what they're trying to accomplish, how they'll measure success, and which group must adapt for the greatest impact.

Next, you'll conduct a *needs assessment* to define the gap between what people are doing now and what you'd like them to do in service of the goal. This difference—called the *performance gap*—will reveal obstacles between current and desired behaviors. You can use these insights to design solutions.

What does a needs assessment entail?

Practically speaking, you'll identify relevant groups within an organization, then use interviews, surveys, and direct observation to answer the following questions.

1. How are employees currently performing, for better or worse?
2. Which behaviors are desirable, and what must change or improve to move towards the business goal?
3. Are there obstacles related to environment, motivation, resources, knowledge, or skills?

Questions one and two offer insights about the boundaries of your performance gap—where are people starting, and where do you want them to end up? Question three reveals what's in the way.

For convenience, I'll refer to learners in this chapter as employees, but these methods also apply to non-business cases.

Who should you ask, involve, or observe?

The first step in a needs assessment is to decide where to gather information. Several groups are worth considering: SMEs, clients, stakeholders, high, low, and average-performing employees, managers, and customers. Each group is valuable, but you can involve them in different ways.

For example, managers may have a clear vision of how they want employees to work, which they can explain directly through interviews or questionnaires. Conversely, low-performing employees might not know what they're doing wrong, so you might need to observe and compare their behavior to others.

Use the methods below to help plan your assessment for each group.

- Ask your clients for existing surveys, analytics, or feedback. They came to you because they discovered a problem—what data were they looking at?
- Interview your SMEs and stakeholders to probe for issues, causes, and successes.
- Speak with managers for their perspectives and their help sourcing employees for review.
- Observe and interview high-performing employees to pinpoint desirable actions and behaviors.
- Observe and interview average performers to set a baseline for what most employees are doing.
- Observe low performers and note how their behavior differs from high performers. Look for trouble spots and obstacles.
- Create digital surveys when there's a large or distributed employee pool, or when you feel privacy is an issue.

- Interview sales or service staff to ask about customer feedback. When appropriate, you might survey customers directly.

After identifying behaviors that define good performance, you can gather high performers, managers, and clients for a review. These mixed-audience discussions can help validate your findings.

Question fodder

While planning your interviews and surveys, ask questions about the causes of poor performance, behaviors you want employees to emulate, and resources they might need. You'll find a few thought starters below.

- What are employees currently doing, and what's wrong with it?
- What are the goals for your employees? How would you like them to perform?
- If they are missing specific targets, why? Look for root causes here: issues in their environment, missing skills or resources, or lackluster motivation.
- What skills and knowledge do employees currently have?
- What skills and knowledge must employees develop?
- What resources (e.g., materials, equipment, people) could help employees perform better?
- What obstacles (e.g., behavior, environment, motivation, knowledge, or skill) might get in the way?

Share your results

Before moving on, summarize and share your results with the client. Consider categorizing insights based on the earlier questions: how are employees currently performing, which behaviors do they need to improve to move towards the business goal, and what obstacles must you help them overcome?

Walking the client through your findings builds a foundation for the design process. As you present work, they'll better understand why you chose one route over another. But first, you should decide which solutions suit the issues you uncovered.

What kinds of solutions will help?

As an instructional designer, part of your role is to identify and address performance issues. But not all problems can be solved with training.

Some issues benefit from simple *job aids*, like lists or on-the-job references. These are useful for tasks with few steps that don't require memorization or practice. For example, if sandwich shop employees can't remember the fixings in a signature hoagie, posting recipes near each station may be all you need! Similarly, if issues arise from company culture, environment, or a lack of motivation, training may not be worthwhile.

Training is most valuable when issues stem from a lack of practicable skills. And when considering eLearning as a delivery mechanism, make sure your tech approach can mimic the conditions you need.

In short, before assuming that training will improve matters, review your needs analysis. Can problems be solved with job aids, or do they require in-depth understanding and practice?

Do issues stem from environmental factors, or are they based on skill?

Once you decide training is viable, your needs analysis should supply a wealth of inspiration for next steps.

A short, sweet example

Imagine a bakery known for its signature custard pastries. After decades of success, the owners suddenly find sales plummeting over the course of three months. They realize this dip coincided precisely with a batch of new bakery hires and suspect a training hiccup. Since the issue impacts their profits and long-standing reputation, the business asks for your help.

Online surveys reveal a slew of customer complaints over the same period. People claim the quality of the famous delicacies changed dramatically, citing inconsistencies in taste, texture, visual presentation, custard volume, and size.

Managers are confident the quality issues stem from the new staff—the timing is too perfect, and there have been no changes in equipment or environment. Further, all complaints arose during shifts when the new bakers worked unsupervised.

As usual, you start by helping the client define a business goal. Since bakery items are tracked by category, sales from the custard pastries seem like a logical metric for success. You work with the client to develop the following:

Increase custard pastry sales by 25% within six months after new bakers complete updated training.

You're not yet sure what new bakers will do to improve, but that's alright. You can refine the statement along the way. With the business goal as your North Star, you start planning your needs assessment.

Review and update existing surveys

Currently, most complaints arrive from an online customer survey. However, responses are limited to star ratings with little room for detail. You update the form with targeted questions and add longer text fields to help customers elaborate.

Interview SMEs and stakeholders

You schedule interviews with bakery managers, counter staff, owners, and the head pastry chef. This process began while developing the business goal and will continue through your needs assessment. You ask managers to help select candidates for interviews and observation.

Interview and observe high performers

You interview veteran pastry chefs, asking them to walk through the age-old recipe and describe their methods. You ask about common pitfalls and how they learned to make the signature pastries. You observe them directly over several shifts and record the process for later. For the good of the project, you even taste-test a few samples. Oh, the sacrifices you make!

Observe low performers

You get permission to observe some of the low-performing bakers directly. You set up in the kitchen when this group works unsupervised, recording their steps and noting differences in technique.

Gather data from sales staff and customers

You interview sales staff at the front counter to ask about customer feedback—some people complain in person or don't bother with online surveys. Verbal feedback is often more detailed.

Assess obstacles and challenging tasks

Once you isolate the issues, you write employee questionnaires to learn more about challenging tasks—especially those that are hardest and most frequent.

Through observation and comparison, you form a picture of why new bakers aren't doing so well.

- The original, decades-old recipe is hugely complicated —the slightest miscalculation in temperature, timing, or measurement results in disastrous pastries. Furthermore, several steps are missing from the written recipe. Veterans rely on expertise and intuition to fine-tune each batch.
- Different batch sizes are produced depending on the time and day. Mornings require piles of pastries, afternoons bring few orders, and weekends fluctuate wildly. Unfortunately, recipe increases aren't linear! When doubling a batch, bakers can't just double the ingredients—veterans calculate a specific formula on the fly. Novices have less practice and default to multiplying ingredients in times of stress. This results in inconsistent quality, especially as batches grow.

- While the pastries are known for their deliciousness, they're also famous for aesthetic appeal. Experienced bakers weave the dough into patterns, each adding unique, artistic flair. Customers grew to expect this level of craftsmanship, and none of this was explained to new hires. Aesthetically speaking, pastries from the newbies are blank slates.
- Finally, the machine responsible for custard mixing (a.k.a. Old Goliath) is a complex and finicky antique. Veterans are accustomed to the quirky dinosaur, but documentation is nonexistent. Whenever the machine malfunctions, new bakers use old, leftover custard or skimp on the filling.

As a result of these issues, new bakers produce inconsistent, inferior pastries, taking twice the time and inaccurately modifying the recipe. Yikes!

While the needs assessment technically ends here, you already see several ways to improve performance. You could...

- ...update the written recipe so it's accurate, easier to learn, and requires less intuition.
- ...create practice activities for recalculating ingredients as batch sizes change.
- ...design dough weaving sessions for new bakers. Since the process is personalized and open to creativity, it may require a high level of practice and skill.
- ...create training to help employees use the old custard machine or suggest purchasing new equipment. If repairs are straightforward and budgets are limited, you might post instructions near Old Goliath herself.

What happens next?

Hopefully, this example offers a short, sweet taste of the needs assessment. You defined the edges of a performance gap, revealed pain points, and gathered insights for solutions.

Now you can share findings with your clients and update your business goal if necessary. You have all the ingredients (sorry, I couldn't resist) to break down each task as needed. And that's what we'll look at next!

Conducting a task analysis

Once you've identified a need, you can design training to address it. You'll break skills into discrete, actionable steps (or tasks) so learners can follow them. Then you'll prioritize the most impactful steps and decide where to focus your training. After you have these steps, you can create instructional activities that suit your audience's abilities and learning preferences. In a nutshell, this is the basis for *task analysis*.

You'll rely heavily on interviews and observation to gather this information, with SMEs and employees as your best sources.

Here's the general workflow:

Start by describing the skill you want to analyze. Decide if it has components or steps that deserve more detail. For example, "De-escalate customer confrontations" is a handy, high-level skill for many jobs, but you'll need to break that down tactically if you want someone to learn it.

Collect all the steps to accomplish a goal by talking to SMEs and learners. Interviewing top performers shows you the route to a successful outcome; watching poor performers reveals gaps you'll want to highlight. Make sure each task is measurable—

you're in the right ballpark if it can be quantified and checked off a list.

Next, prioritize what's most relevant for your training. Difficult, critical, and frequently-performed tasks are good candidates. You might ask:

- How often is this step performed?
- How difficult is it for your target audience?
- What knowledge or abilities does it require?
- How crucial is this step for the primary goal?
- What are the consequences if this step is omitted or done wrong?

As instructions take shape, validate your steps with SMEs and have others follow them to find gaps. Create practice activities and materials based on your insights.

Say you're creating training for a wilderness search-and-rescue team. Topics range from orienteering, to edible local plants, to emergency first aid.

Dealing with sprains and broken bones is one critical skill for this group—patching up injured hikers so they can walk themselves to safety. But addressing injuries in the wild is no easy business. Rescuers must distinguish between a bruise, sprain, or fracture, decide whether a person can be moved, and possibly set and splint the bone. They'll assess medical supplies, consider the weather and terrain, and perform dozens of steps between arriving at a scene and getting their patient home.

Running a complex process through task analysis makes your job more manageable. It helps you locate obstacles, validate instructions, and create activities that match the skills of your audience. And while you may only conduct a needs assessment once, task analyses can be performed repeatedly for each new idea.

Learner analysis

Before customizing material for a group, you'll want to know as much as possible about them. During *learner analysis*, you'll examine learners' skills, knowledge, and preferences to tailor instructional content.

This is another way of saying you should learn about your learners, and we've covered several methods for doing so. You might interview SMEs and learners directly or cast a wide net with focus groups. Surveys are also helpful when dealing with a large workforce.

While learner analysis applies to any ID project, it's especially relevant for eLearning. Not only must you assess what learners know, you need to ensure they're comfortable with the tech you're using. There's no point in hosting a pool party when your friends can't swim!

So, in addition to asking questions about their existing knowledge, you'll want to gauge technical savviness and delivery preferences.

Does your audience learn best from video webinars, interactive experiences, or live group sessions? What devices do they use most frequently, and what's their comfort level? Do they run screaming from technology and prefer face-to-face instruction?

If you find a group with mixed expertise, you still have options. You might divide them into separate classes, design for your least savvy learners, or include more guidance in your experience.

Context analysis

After establishing your learners' abilities and knowledge, it helps to consider what's around them. During *context analysis*, you'll study the settings that impact your learners either during training (the *learning context*) or in the real world (the *performance context*). You can gather this information through interviews or by inspecting each context directly.

Where will your training take place, and what's the environment like? Are there distractions, conditions, or deprecated technology you need to deal with? Is your eLearning remote?

Observations like these can shape your material and help you plan for quirks, hiccups, or hurdles in the training environment. If a death metal band plays outside your office on Fridays, you may want to schedule around that!

And yes, in eLearning, the definition of setting or environment depends on the project. But the closer you plan for reality, the smoother your learning experience.

Do I have to use all of these methods?

Heck no! You don't have to do anything. And there's certainly overlap here. Some people combine tactics, while others rely mostly on action mapping or learning objectives (up next) and supplement as needed. My goal is to make you aware of each method, its purpose, and how to use it.

In the next chapter, you'll get a better grasp of how everything fits together.

Chapter takeaways

By deciding what learners must do and dissecting each step, instructional designers can build eLearning to fill performance gaps. In this chapter, we looked at ways to analyze your audience and shape your game plan.

- Start with a *needs assessment* to gauge your learners' abilities and goals. Then, use *task analysis* to break down the steps in between.
- Conduct a *learner analysis* to find strengths, weaknesses, and preferences, and a *context analysis* to understand conditions during and after training.
- By incorporating these insights, you can tailor content to specific needs and prepare learners for reality.

10

BUILD OBJECTIVES, ACTIONS, AND PRACTICE ACTIVITIES

In the last few chapters, we identified our SMEs, learned tips for interviewing them, and dissected needs and tasks. Now, let's go further down that road and find ways to plan our content.

We'll begin with two questions:

- Based on your business goal, what should learners be able to do after they finish your lessons or training?
- How will you structure your material to help them get there?

With *Bloom's Taxonomy*, you'll target a level of knowledge learners need to reach, then write learning objectives to match. Based on those objectives, you'll create materials and decide how to measure progress.

Alternately, Cathy Moore's *action mapping* is a technique for diagramming tasks used to accomplish a goal. Here, you'll analyze a performance issue, draw a map of actions learners can take to address it, then design practice activities.

While action mapping is more recent, both methods are still used in modern instructional design. And in most cases, you'll choose either one or the other.

Learning objectives

While a business goal outlines what an organization wants to achieve, a good *learning objective* describes what people should be able to do.

Let's say a coffee shop receives low ratings due to a lack of beverage options. They develop the following business goal:

Increase online ratings to five stars by the end of Q3 after baristas expand their drink-mixing repertoire.

As a result, the owners purchase a next-gen espresso machine, but the thing has sixteen modes and can mix three hundred specialty drinks! With a complexity rivaling the lunar shuttle, the need for training is evident.

A well-written learning objective needs a few key ingredients. First, it should have a practical way to gauge success.

If I wrote: "Learners will understand the new espresso machine." Well, that's a fuzzy objective for a few reasons. Most notably, the topic is too broad. What features do learners need to know? Are they just turning this thing on and off, learning about hot foam, or should their knowledge go deeper? And how would you measure successful understanding?

Your objective should also have an impact on the business goal. In other words: you're designing training to accomplish something and fill gaps in knowledge or skills, right? Your learning objective should directly narrow those gaps.

Generally, you can use the following format to ensure your learning objective has the basic ingredients:

By the end of [your learning experience], learners will be able to [verb + measurable objective here].

Customizing it with the previous example, you might have something like:

By the end of training, baristas will be able to operate the machine to mix thirteen new espresso drinks.

In this example, the timing is clear (by the end of training), measures for success are defined (operating the machine to mix thirteen new drinks), and the outcomes impact the primary goal (employees learn how to use equipment to expand the drink menu).

Of course, you can add detail as needed—maybe the manager's responsible for judging whether drinks are mixed properly. But this statement addresses the business goal, and there's a way to gauge success.

You may have noticed that a chunk of what makes learning objectives strong or weak, measurable or vague, comes from the verb you choose. Verifying whether learners "understand" something is ambiguous—you'd always need to write more to clarify what that entails. Using actionable verbs can focus your objectives and make them easier to measure.

Maybe there's a method to help us with that. Hint, hint.

Better objectives with Bloom's Taxonomy

Now that we've got some context, let's talk about *Bloom's Taxonomy*.

As in the barista example, it's helpful to look at learning goals from a few angles.

- How will people use their new knowledge? In other words, what kinds of tasks, interpretations, or activities will they have to do?
- How deeply must people learn the material to do these things, and how much effort should they invest?

Benjamin Bloom developed his taxonomy to classify learning by type and depth, and to write learning objectives with suitable verbs (Bloom, 1984). In other words, you'll answer the questions above, look at Bloom's for options, then choose a corresponding verb for your statement.

The process has been tweaked over the years, with more levels added in the 1990s and phrasing changed from adjectives to verbs (Anderson & Krathwohl, 2001). In this book, we'll use a version with six levels of learning, phrased as verbs and arranged in a hierarchy (Armstrong, 2010).

IN THIS TIER:	LEARNERS SHOULD BE ABLE TO:*
CREATE	**PRODUCE NEW OR ORIGINAL WORK** **verbs:** *design, assemble, construct, conjecture, develop, formulate, author, investigate*
EVALUATE	**JUSTIFY A STAND OR DECISION** **verbs:** appraise, argue, defend, judge, select, support, value, critique, weigh
ANALYZE	**DRAW CONNECTIONS AMONG IDEAS** **verbs:** differentiate, organize, relate, compare, contrast, distinguish, examine, experiment, question, test
APPLY	**USE INFORMATION IN NEW SITUATIONS** **verbs:** execute, implement, solve, use, demonstrate, interpret, operate, schedule, sketch
UNDERSTAND	**EXPLAIN IDEAS OR CONCEPTS** **verbs:** classify, describe, discuss, explain, identify, locate, recognize, report, select, translate
REMEMBER	**RECALL FACTS AND BASIC CONCEPTS** **verbs:** define, duplicate, list, memorize, repeat, state

INCREASED EFFORT & DEPTH OF LEARNING (arrow pointing upward from REMEMBER to CREATE)

* Based on a diagram from the Vanderbilt University Center for Teaching. Creative Commons Attribution license. (Armstrong, 2010)

The lowest level, *Remember*, only requires learners to retain what they've been taught. They don't need to do much with it yet. But as you move from the bottom to the top, the depth of learning, complexity, and effort increases. You'll often see this referenced as lower-order or higher-order thinking.

Here's an outline of what learners should be able to do at each tier of Bloom's:

- **Remember:** Recall information. This is the lowest level of learning.
- **Understand:** Absorb information well enough to explain it.
- **Apply:** Use information in unpredictable situations that differ from the learning environment.
- **Analyze:** Make logical connections, find relationships, and draw conclusions.
- **Evaluate:** Formulate opinions or judgments based on interpretation.
- **Create:** Use learned information to produce something new.

Each level builds upon the previous one. To understand a concept, you must first remember what you learned about it. Before you can apply what you've learned, you must understand it. And so on.

When writing learning objectives for your specific project, what are you asking of your learners? Do you just need them to remember something? Say, to rattle off simple instructions or recite a few steps? Do they need to apply information? Or should they grasp something so deeply that they can interpret it creatively to make something new?

These answers are crucial for design, because building deep knowledge requires effort from both you and the audience. For example, it may take minutes to memorize a recipe, but years to master creative cooking. And you don't want to build Sous Chef training when someone just needs to scramble an egg!

To clarify: the six buckets from Bloom's Taxonomy (Remember, Understand, etc.) are titles for different levels of learning. They're not the actual verbs you'll use in your statements. Each level contains a group of verbs to make writing objectives easier. In other words, you wouldn't use "apply" as a word in your statement—that would be the level of learning you're hoping to achieve. Instead, you'd write, "By the end of the experience, learners will be able to *schedule* meetings using the new system." Because schedule is a verb from the Apply tier.

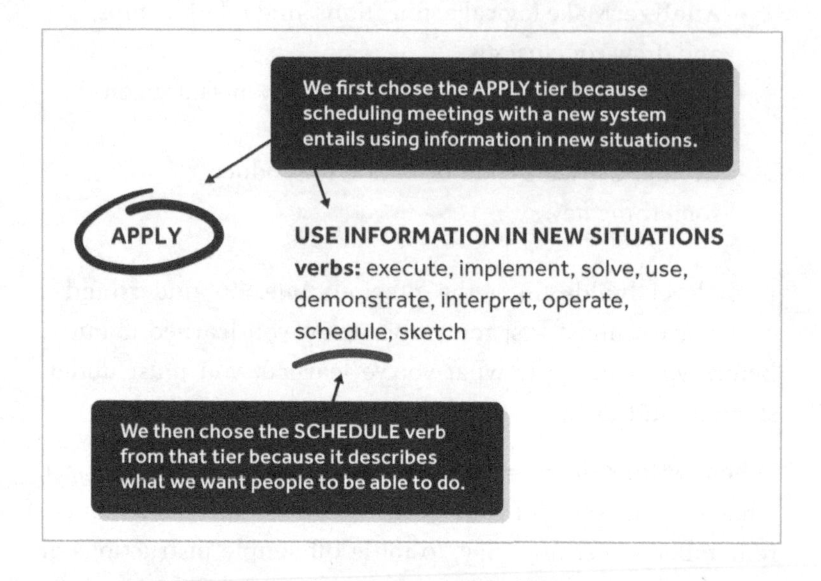

Yup, I know. That's where people get confused. But with that in mind, the steps for writing learning objectives using Bloom's Taxonomy are as follows:

1. Decide on your instructional topic or material.
2. Articulate what you want learners to be able to do with this knowledge after completing your lesson or training. Be specific. Do they need to remember information, apply it in new situations, or create something new?

3. Reference Bloom's Taxonomy with this in mind. Consider how deeply your audience needs to learn the material to use it properly. Decide which tier implies the level of connection, complexity, and investment required from your audience.
4. Look at the verbs within that tier and, if necessary, visit *Tools & Resources* for more synonyms. Pick a verb that aligns with what learners must do, and use it in your learning objective.
5. Add a specific, quantifiable task or metric to help measure success. The easier it is to check off, the more valuable it will be for evaluation.
6. You should now have a statement like: "*By the end of the experience, learners will be able to [verb + measurable objective here].*"
7. Using your shiny new learning objective as a guide, brainstorm activities and content that help learners reach that level of understanding and target your measurable goal.

Whew! Still with me?

Once you've drafted a clear learning objective, you'll use it as guidance while designing materials and activities. Does a lesson move your learner closer to your goal? Does a practice activity directly steer them in the right direction? After learners get through all your content, you now have a way to measure whether you've succeeded.

While it may take practice, Bloom's Taxonomy is useful for considering depth and complexity in your learning objectives. And if all that was as exhausting to learn as it was to write, you've earned yourself a margarita break!

After that, we'll move on to action mapping.

Action mapping

Action mapping is a more recent addition to the instructional designer's toolkit. It was created by Cathy Moore (2017) and detailed in her book, *Map It: The hands-on guide to strategic training design*.

Action mapping is an all-in-one visual exercise for developing learning materials. It includes elements of needs assessments and task analysis, and is often considered an alternative to learning objectives. It's most useful for solving performance-related issues, less so for academic projects that promote learning for its own sake.

With this approach, you'll work with stakeholders to capture actions and behaviors that contribute to a business goal. You'll then design practice activities based on these actions.

Here's (roughly) how it goes.

1. Prepare a blank canvas.

You thought I was kidding back in the digital tools section? Nope! Here we are. Spin up your favorite mapping app, a digital or physical whiteboard, or a fat stack of sticky notes. Action mapping is a visual exercise, so you need somewhere to collect and reshuffle ideas where everyone can see them.

Why everyone? Because you'll want to conduct these first steps with your SMEs and stakeholders to ensure they're all working towards a consensus. Not only does it get them involved, but these groups help define issues, goals, and actions.

For a close look at the maps from this section, jump to *Tools & Resources* for links.

2. Define a measurable business goal.

We discussed this in the *Work with SMEs and stakeholders* chapter, and the process is the same. Writing a business goal is crucial. But getting clients to commit to a central issue—and agree on a metric for success—can be challenging.

Let's say a client arrives with a remarkable product line: supplements that grant various superpowers, from laser vision to flight! Unfortunately, the business is facing a high volume of customer returns and enlists your help.

Stakeholders think their support team plays a key role. Customers contact them over chat, phone, or email before requesting a refund, and improving this group's performance could reduce the return rate. With this in mind, you and your client begin with a clear business goal:

Reduce supplement returns by 15% in six months after support staff employ new response tactics.

You place this in the center of your action map.

Reduce supplement returns by 15% in six months after support staff employ new response tactics

FYI: Maps in this book were created with MindMeister.

Great! You've defined the finish line, but that's just the beginning. Next, you'll determine what people can do to solve the problem.

3. List actions people can take to work toward the goal.

As you continue this exercise with your group, ask: what issues exist, and what can people do to address them? Your SMEs and stakeholders offer several reasons for frequent returns.

- **Poor recommendations:** Customers complain that staff recommendations don't align with their goals. For instance, shoppers voicing their love of travel are often sold pills for super speed. However, most return these products after realizing teleportation would have been a far better fit.
- **Failure to anticipate adverse interactions:** Customers report negative interactions with their existing medications or supplements and feel customer support should have warned them. People experience sudden teleportation, pyrokinesis, and elasticity in all the wrong places, resulting in lawsuits galore.
- **Failure to offer exchanges:** Support staff fails to provide alternatives to product returns. An exchange program exists, but employees are unfamiliar with the details.
- **Customers are confused by similar products:** The product lineup contains similar items that are poorly explained by support staff. Confused by dosages and packaging, customers often purchase the wrong supplement and return it later.
- **Trouble accessing profile and purchase history:** Customers complain that support has trouble

accessing their profile history on chats or calls, resulting in frustration and repeated explanations.

Next, you'll translate these issues into actions: what can support staff do to improve? Actions should be specific, tactical, and easily checked. As with learning objectives, be sure to watch your verbs—calls to "define" or "understand" are hard to measure.

As your SMEs throw out ideas, capture them as children of your main goal. If you can split an action into smaller steps, add those as well.

In our example, the actions and sub-actions might include:

> **ACTION:** Recommend products that align with customer goals.

- Ask diagnostic questions to assess needs and goals.
- Choose fitting supplements from the product catalog.
- Discuss each supplement's benefits and verify a good match with the customer before purchase.

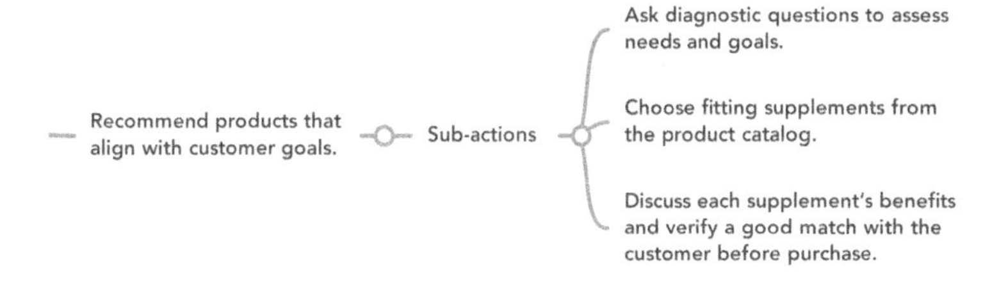

ACTION: Identify drug or supplement interactions, then alert customers before purchase.

- Check the customer's profile for existing medications and supplements.
- Ask the customer about their current regimen. Verify against their profile and update as necessary.
- Cross-reference the customer's regimen against known product interactions.
- Alert the customer to possible issues, then suggest remedies or refer them to medical professionals.

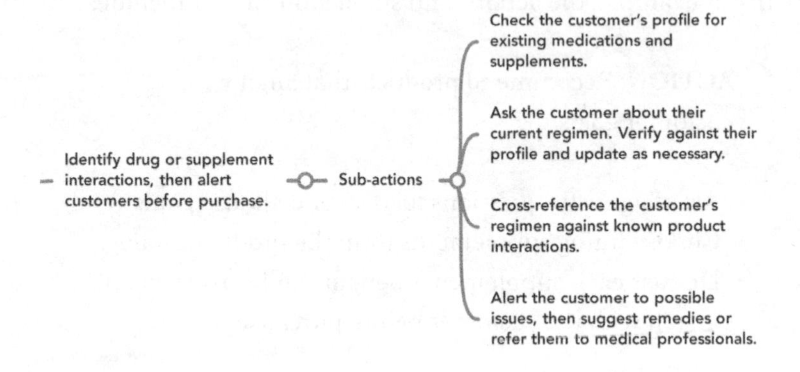

ACTION: Offer exchanges and alternatives to product returns when appropriate.

ACTION: Answer customer questions to distinguish between similar products.

ACTION: Access customer profiles quickly to minimize frustration and repetition.

As your map takes shape and you reach a healthy threshold, start prioritizing actions with the highest impact on your main goal. When performing a task poorly has extreme consequences, you might bump it up the list.

Oh, and if you see some overlap with needs assessment and task mapping, you've been paying attention! While action mapping is much more regimented, these methods share common ground with identifying, prioritizing, and dissecting what people must do.

4. Brainstorm activities to practice these actions.

After defining a goal and the steps to get there, it's time to design practice activities! For each action in your map, brainstorm exercises to help learners build their skills. You can use scenarios to present a challenge, drive decisions, and reveal the consequences. This active learning improves retention through feedback and critique.

In this case, you'll devise ways to help the support staff practice their customer interactions.

- Interactive videos to choose appropriate supplements based on customer conversations and goals. *(eLearning)*
- Scenarios to practice questioning customers, identifying adverse effects based on medications and supplements, and responding accordingly. *(eLearning)*
- Scenarios to identify when exchanges apply and practice offering options. *(eLearning)*
- Roleplay with a manager posing as a customer confused by the product lineup. *(On the job)*
- Real-time practice locating profiles with the existing customer service system. *(On the job)*
- Upgrade the customer service system to automatically identify customers by email or call-in number. *(Tools)*

On your map, this may look like:

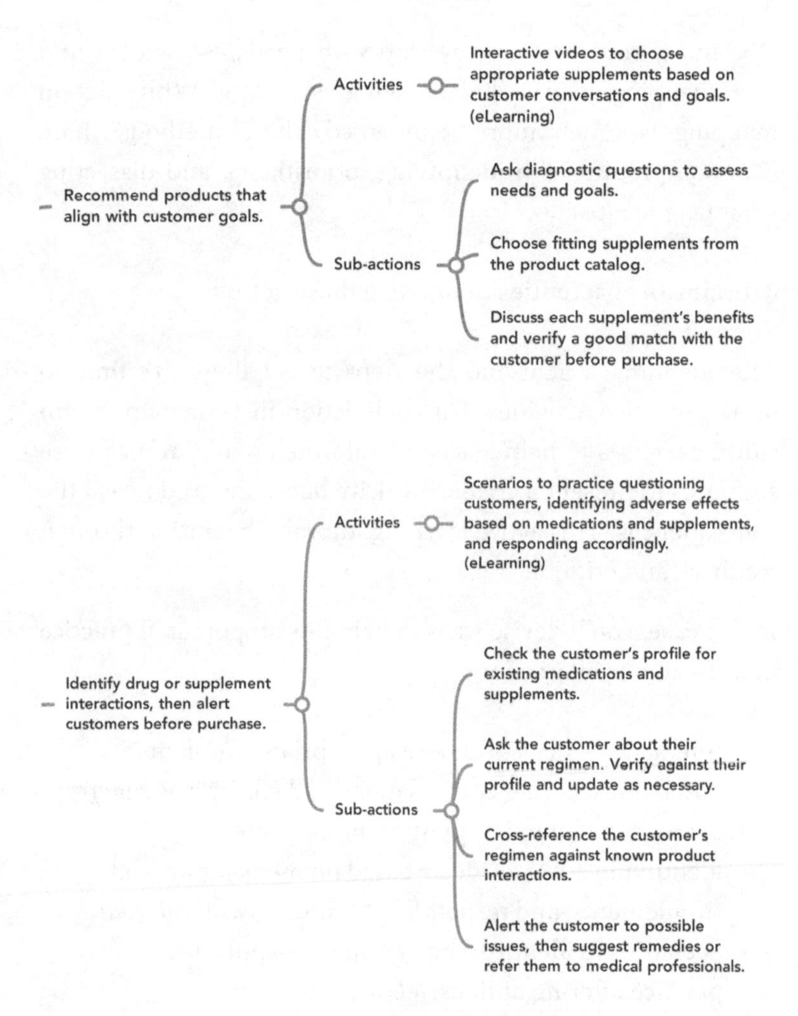

Remember, not all problems require training or eLearning. That's okay! You're considering all viable solutions at this stage. You might solve some issues through job aids or in-person training; others may benefit from tools or software. Don't limit your thinking yet.

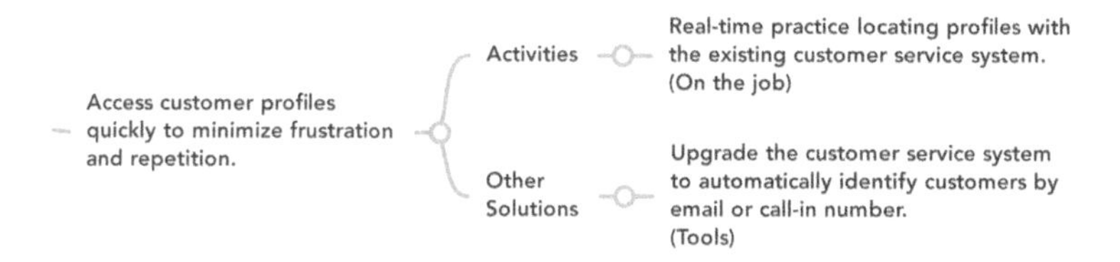

5. Distill the required information for each activity, then teach it in context.

Things are taking shape! Now you can review your activities and add the minimum information a learner would need to accomplish them. Rather than simply telling people what they need to know, you'll reveal this information in context as a learner makes decisions.

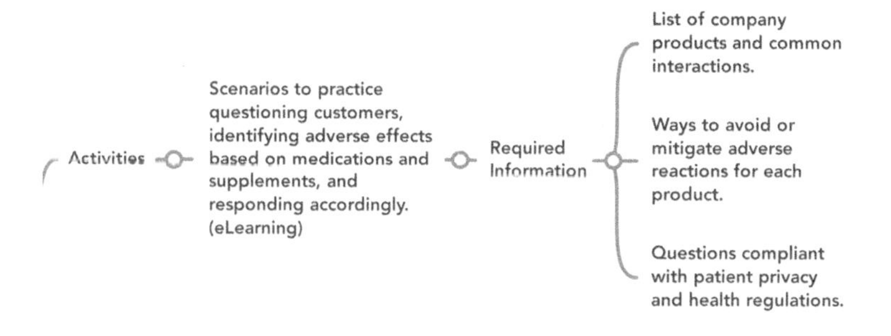

Let's say you want learners to practice questioning customers, identifying adverse effects based on current medications or supplements, and responding accordingly. To do this, support staff needs the following information.

- A list of company products and interactions with common drugs or supplements.

- Ways to avoid or mitigate adverse reactions for each product.
- Which questions are compliant with patient privacy and health regulations.

But how can you work this information into your exercises?

Method 1: Telling and quizzing

In a traditional setup, you might list some data points and follow up with a quiz. For example:

Which of the following may cause unexpected pyrokinesis when combined with our products?

1. *Vitamin D*
2. *Vitamin B*
3. *Omega-3 fatty acids*

The correct answer, of course, is Vitamin B. If someone chooses that, you tell them so.

Thrilling, yes?

Quizzes can be useful for short lists or limited retention, but they don't make the brain work very hard. Less effort builds weaker memories.

Method 2: Scenarios and feedback

For more active learning, try writing scenarios with insights in your feedback. When a learner chooses a favorable outcome, you reinforce and elaborate. If they pursue the wrong road, show them the consequences. For example:

A customer named Mica calls with a concern. She's attending a pool party and wants something to help her breathe underwater, but her daily regimen includes vitamin D, folic acid, and fish oil. Will this be an issue with the aqua lungs product?

1. *Yes. Vitamin D interacts negatively with products for all water-based superpowers.*
2. *She should be fine. None of her supplements conflict with the aqua lungs product.*
3. *When taking aqua lungs, she should eat asparagus each morning to avoid issues with her regimen.*

When your learner chooses option B, they get the following result:

Two days later, Mica calls you in a fury. Apparently, she burst into flames at her pool party and evaporated all the water! Her online search revealed that folic acid is a form of vitamin B—a clear no-no for the aqua lungs protocol. Angry and humiliated, she demands a full refund.

Which example better prepares learners for their jobs? Which helps them practice and interpret information?

Also, if you need a hint: asparagus is great for all kinds of things. So much fiber.

6. Keep rolling!

You now have an action map showing the relationship between your business goal, actions your learners can take, exercises to help them practice, and information they need. This logical chain helps prune and protect your eLearning content against extraneous material.

We'll use all this in the next chapter as you start outlining.

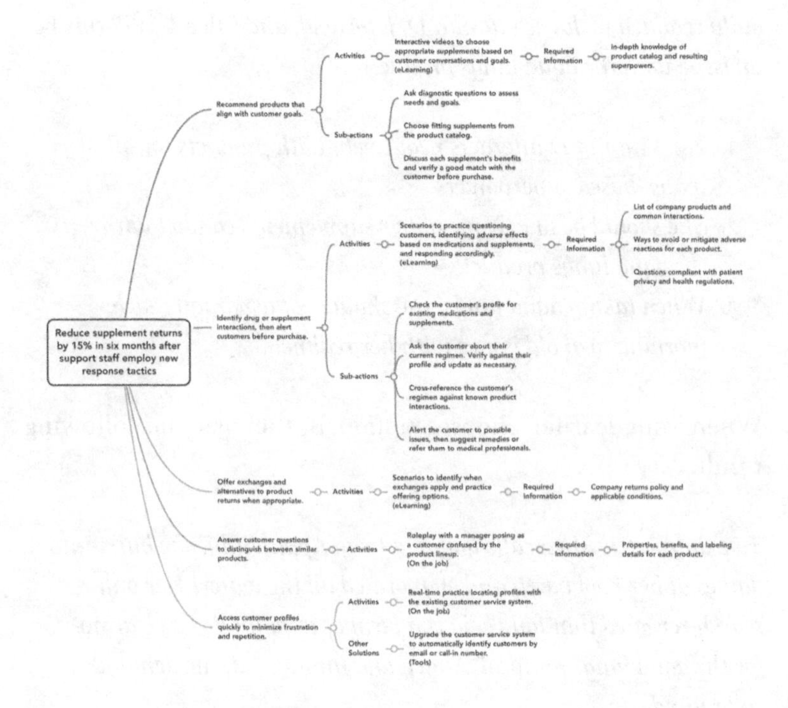

No, this diagram's not small. You just have tiny eyeballs. You'll find links to a full view in Tools & Resources.

Action mapping is worth a deeper dive. For more nitty-gritty details, I recommend following Cathy Moore online and reading her book *Map It: The hands-on guide to strategic training design*.

Common questions:

If I choose action mapping, do I still use other methods?

According to Cathy Moore, "Action mapping includes needs analysis; it's not just a way to organize content." (Moore, n.d.) However, some instructional designers conduct a dedicated needs assessment, then move on to action mapping for practice activities.

A similar logic applies to task analysis. During action mapping, you dissect actions and sub-actions, then prioritize critical steps. Well, that sounds a lot like task analysis, doesn't it?

Ultimately, you now know what each method is meant to do. If your action mapping thoroughly defines a performance gap and breaks down each step, you may not require a needs assessment or task analysis.

When should I schedule action mapping in my workflow?

Schedule action mapping sessions with SMEs and stakeholders early in your project, especially the first steps where you write a business goal. Everyone must agree on that core statement to steer the process and help SMEs gather material.

When should I choose action mapping vs. learning objectives?

Action mapping offers guidance and structure. It involves stakeholders in brainstorming, creates practice activities that impact your goal, and ensures meaningful content. If you're addressing performance-based issues and want step-by-step tactics, action mapping may be a good fit.

However, since action mapping relies on behaviors, it's less effective for pure academics or cases where knowledge is the only goal. In fact, required knowledge is the *last* thing you define in action mapping!

Learning objectives also steer the design process, and they've been around for a long time. You still develop a business goal, but beyond that, the approach is less prescriptive.

Once you write learning objectives, you'll need other tactics to build activities and content. This method can work well for academic projects or if you prefer a less rigid approach. Objectives can also help summarize learning benefits for an audience.

Chapter takeaways

What should learners be able to do after training, and how can you design materials in service of these goals? In this chapter, we discussed two ways to answer these questions and took a brief walking tour.

- Take a deep dive into learner goals and your strategy for meeting them using either *Bloom's Taxonomy* or *action mapping*.
- In Bloom's Taxonomy, you'll decide how people will use knowledge and choose the depth of learning required. Then you'll write a *learning objective* to guide your work.
- With action mapping, you'll diagram all the actions contributing to a goal, then design practice activities. This approach includes elements of both needs assessment and task analysis, and serves as an alternative to learning objectives.

11

CREATE YOUR OUTLINE

Yeah, I know. You did loads of these back in high school. Well, let's see if we can add a few sprinkles to this sundae, shall we?

Outlines are exactly what they sound like—rough guides for structuring your content. They provide a sense of how learners move between topics and help you find gaps before churning out piles of material. Above all, they'll offer a bird's-eye view of your experience.

In this chapter, we'll see what makes eLearning outlines unique and dissect how to build them. Then we'll explore *Gagné's Nine Events of Instruction* on our way to storyboards.

Before you start

By this point, you understand your audience and what they need to accomplish. You uncovered requirements, interviewed stakeholders, and wrote a clear project goal.

Maybe you conducted a needs assessment and broke down some tasks. If you action mapped, you've arrived with a stash of activities. Regardless, the purpose of an outline is to start organizing content. So before you dig in, make sure you've gathered some!

Types of eLearning structure

eLearning is a kind of storytelling, and there are many ways to build a narrative. You might deliver a story in a straight-up, linear fashion, then ask people about what you've said. Or you might drop the reader in different scenarios to see: "What would you do here?"

The structure of your outline should support your approach. And while your experience needn't be entirely one or the other, there are two common options.

Traditional

Traditional structure is what most people envision when they think of eLearning. In its simplest form, you create goals for your learners, tell them what they need to know, then reinforce lessons with exercises.

Learners move through material sequentially, building their knowledge one block at a time. First, we introduce the stove, then the pan, then the butter and eggs, then the folding technique. Verify knowledge at the end, and voilá—we've outlined the omelet!

This approach is easy to grasp, and is figuratively and literally straightforward. But the classic lecture-and-test format can sometimes glaze the eyeballs.

Scenario-based

Story or *scenario-based* experiences put learners in situations where they make choices. They learn by doing, failing, succeeding, getting feedback, and adjusting. Rather than explicitly stating facts, you'll embed lessons within each scenario.

This method often uses branched storylines, where learners explore diverging paths and consequences. You can probably imagine why diagrams come in handy as you consider all the what-if possibilities.

Scenarios encourage learning through immersion and practice. But even the most scenario-heavy experiences benefit from outlining. You may want breaks or summaries to vary the tempo, review insights, or avoid exhaustion from back-to-back exercises. And at minimum, an outline is useful for sequencing activities.

What makes eLearning outlines unique?

You'll eventually design each slide in your eLearning experience—on-screen text, voice-over, interactivity, etc. But before that, you need to decide on topic order, activities, and how learners will navigate. Outlining is an intermediate step between gathering content and detailing every nook and cranny.

Traditional text outlines focus on sequence and hierarchy. For example, when planning a book or presentation, you string information from beginning to end because that's how people experience them.

However, eLearning can diverge from a fixed path, allowing users to explore topics non-linearly, as one would navigate a website. They may choose various routes or discover multiple outcomes.

With that in mind, there are three things your outline should do.

- Establish a hierarchy of topics, sub-topics, details, and activities.
- Organize content into a logical order.
- Suggest user paths and dependencies.

What does all this look like? Great question!

Formatting your outline

You can format your outline in different ways, depending on your needs and preferences. The first option is to write a simple text hierarchy.

Project title

- **Topic 1**
  - Subtopic
  - Subtopic
  - Activity
- **Topic 2**
  - Subtopic
  - Subtopic
- **Topic 3**
  - Subtopic
  - Subtopic
- **Summary**

This method lets you write in long form and requires only bullets and indents to build your structure. However, when we use plain text for eLearning outlines, it can raise questions about *user flow*. Does someone need to finish Topic 1 before moving to Topic 2, or can they choose either to start? What happens after users complete each activity? Text-based outlines are great for showing hierarchy, but you'll need annotations to explain dependencies or movement.

Branched scenarios can be especially tricky. The more elaborate a decision tree, the harder it is to visualize with text alone. Furthermore, rearranging topics with raw text can be a tedious game of copy-and-paste. And as content grows, it's harder to see the big picture.

As an alternate format, you can create a visual outline using methods similar to action mapping (and again, flip to *Tools & Resources* for a link to larger views).

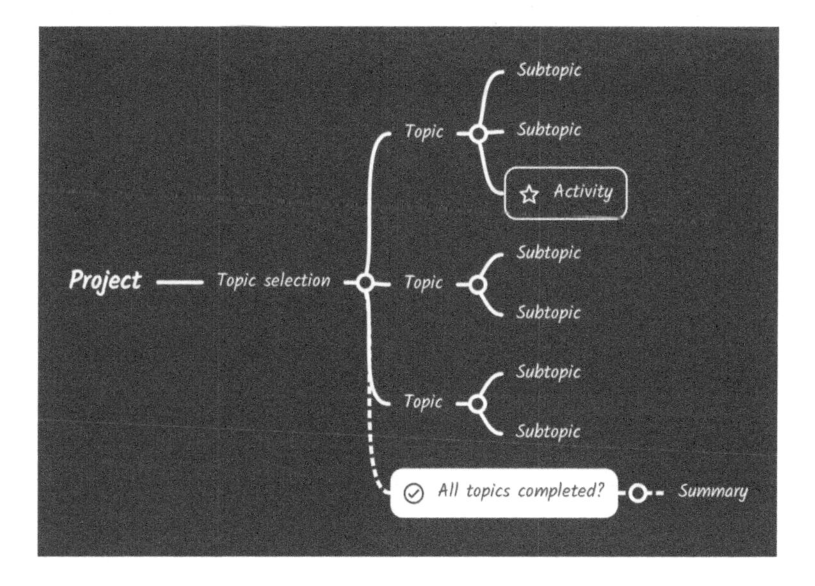

This option is great for quick thinking and a less linear outline. The experience of dragging, dropping, or connecting feels more like building than writing. You can differentiate content with icons or styling and use arrows to specify user flow. Some apps even let you collapse nodes to keep things tidy.

To be fair, visual outlines aren't ideal for lengthy text. They can grow overwhelming when information gets dense and digital tools vary widely. Still, the process feels highly creative, and the benefits are worth a learning curve. If you get frustrated, just revert to bullets and text.

Regardless, the goal of your outline is to organize material for storyboarding. Choose your format, and we'll explore a cosmic example.

Space gardening!

Growing a sustainable space garden is crucial for long missions, but poses a challenge due to zero gravity and limited resources. Astronauts have had difficulty keeping plants alive on recent (fictional) assignments, with a survival rate of only 11%!

After several sessions with SMEs and mission planners, you and your client agree on the following project goal.

> **Increase plant survival rate to 60% by Q4 after astronauts follow the space garden design and maintenance protocol.**

You've generated practice activities through action mapping, but your client wants additional structure. With research and material in hand, you start building an eLearning outline.

Step-by-step workflow

Outlining is a process. You'll begin with an excess, distill critical beats, and restructure along the way. By the end, you should have a scaffold for your material rather than a bloated, illegible mess. Here's a loose formula for bringing it all together.

List topics based on your material

You've done the research, so review your existing content and consider how you might group ideas. Do you see topics or overarching themes to organize your material? Collect these in a document without worrying about the order.

You can use a mapping tool, plain text, or any medium that lets you quickly reshuffle. When using text alone, remember that although topics appear sequential in writing, you haven't yet assigned an order.

A visual outline at this point might look like the following, implying no sequence or structure. In other words, we won't assume which topic appears first, second, or third, if there's an order at all.

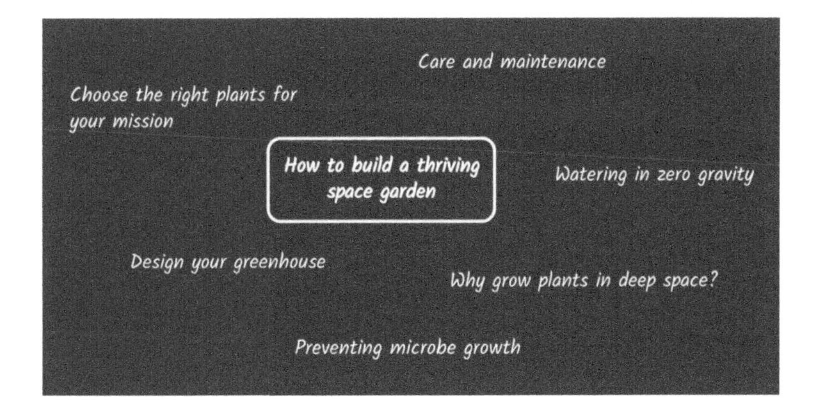

Conversely, a text outline implies a sequence, whether you intend one or not. This certainly isn't a deal breaker; just keep it in mind.

How to build a thriving space garden

- **Why grow plants in deep space?**
- **Choose the right plants for your mission**
- **Watering in zero gravity**
- **Care and maintenance**
- **Preventing microbe growth**
- **Design your greenhouse**

For brevity, we'll show mostly visual outlines from here on. Hopefully, the text equivalent is clear, and we'll revisit it at the end.

Add content and hierarchy

As you add details from your up-front work, some concepts will naturally fit within others. Start grouping topics, subtopics, and core information. Insert activities from action mapping or brainstorms to see where they fit.

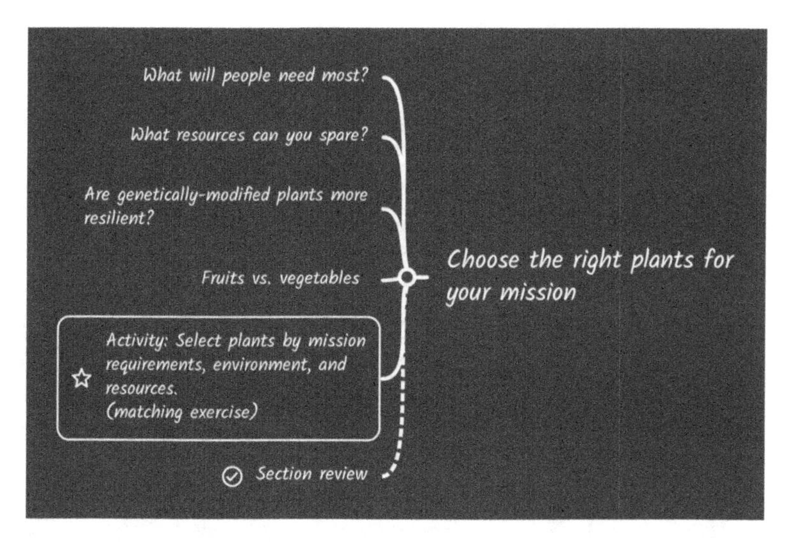

Choosing the right plants emerges as a primary topic.

We'll call each node from your outline a *story beat* for reference, but it needn't correspond to a single eLearning screen. It's a shorthand to avoid distraction.

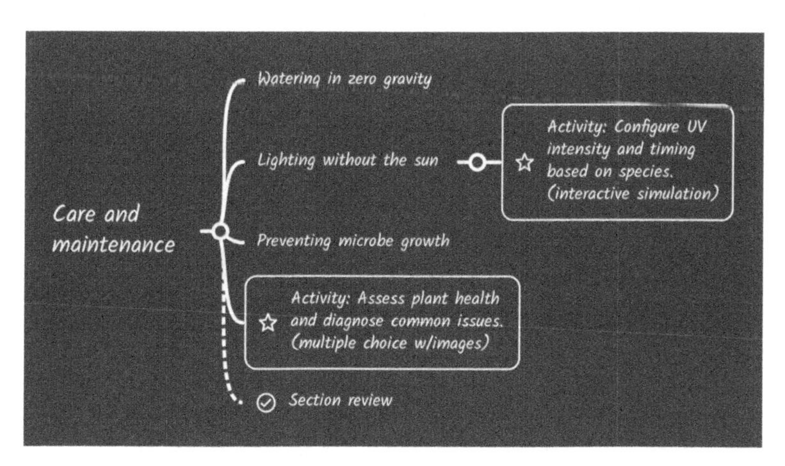

We add two activities and some structure to care and maintenance.

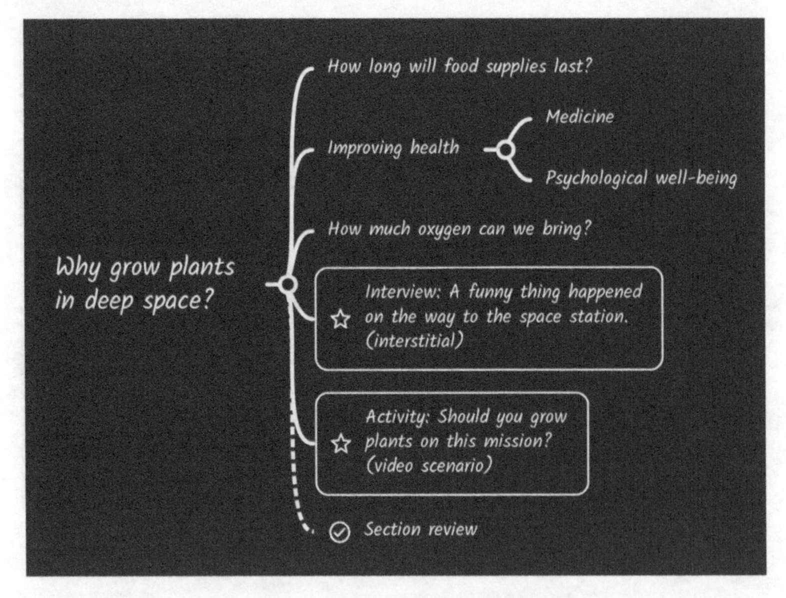

Why grow plants in deep space? So many reasons! You might even branch into subtopics.

In some cases, a string of scenarios might comprise most of your eLearning experience, but you'll still need to put them in order.

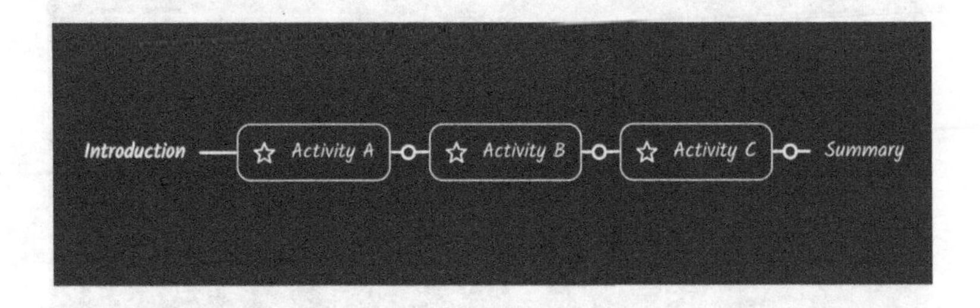

You may also find all kinds of visuals popping into your head for interfaces or UI. Add these as notes in your outline. For example: "This task would work well as a matching game with illustrations for X, Y, Z ." Consider these reminders to your future self when you start designing slides.

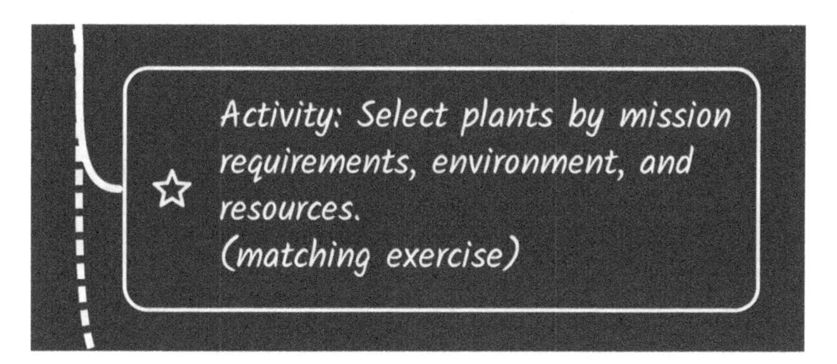

Can you choose plants suited for your mission in a matching exercise?

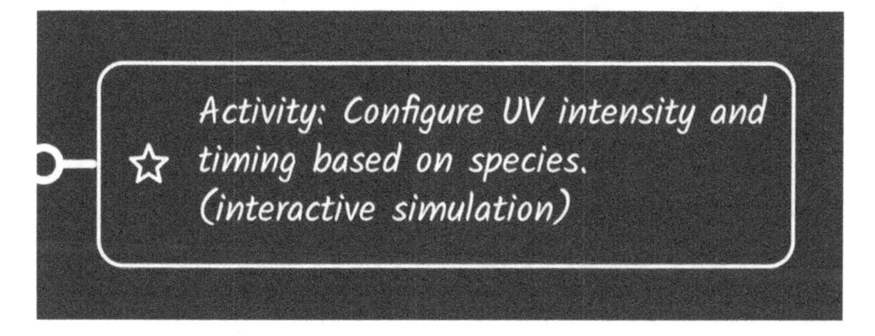

Configure your lighting based on species in this life-or-death simulation.

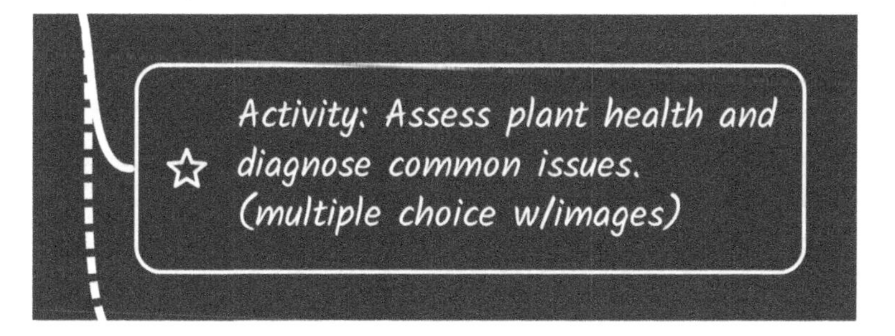

Diagnose plants in distress and decide what to do!

You can add notes and required information from your action mapping as well. This collects everything in one place and makes storyboarding easier in the next step.

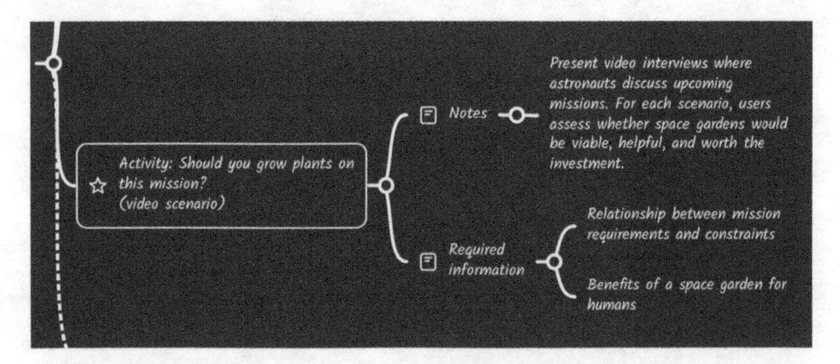

Attach notes and required info as children of your activities. Icons make your outline easier to read at a glance.

Suggest sequence and user flow

For interactive experiences, UX designers often create user-flow diagrams to describe possible paths through a site, app, or eLearning experience. And while that's not the main goal of your outline, you can still consider how learners navigate.

Will you lead users through a linear narrative or let them choose their own route? Will you build decision trees? Perhaps you'll reinforce lessons with previews and summaries.

The outline is your chance to plan a logical progression. Show where choices split and which tasks remain locked until others are completed. Use notes or connective lines to describe movement. Set style conventions that make sense, then use them consistently.

How detailed should this be?

Your outline can be broad or ultra-granular. If you want to plan every scenario, branch, state, and interaction, great! If you just need space to arrange ideas and activities, that's fine too.

Ultimately, your outline provides a high-level content map for each beat in your story. It lets you see where things connect, how users navigate, and the overall scope of your work.

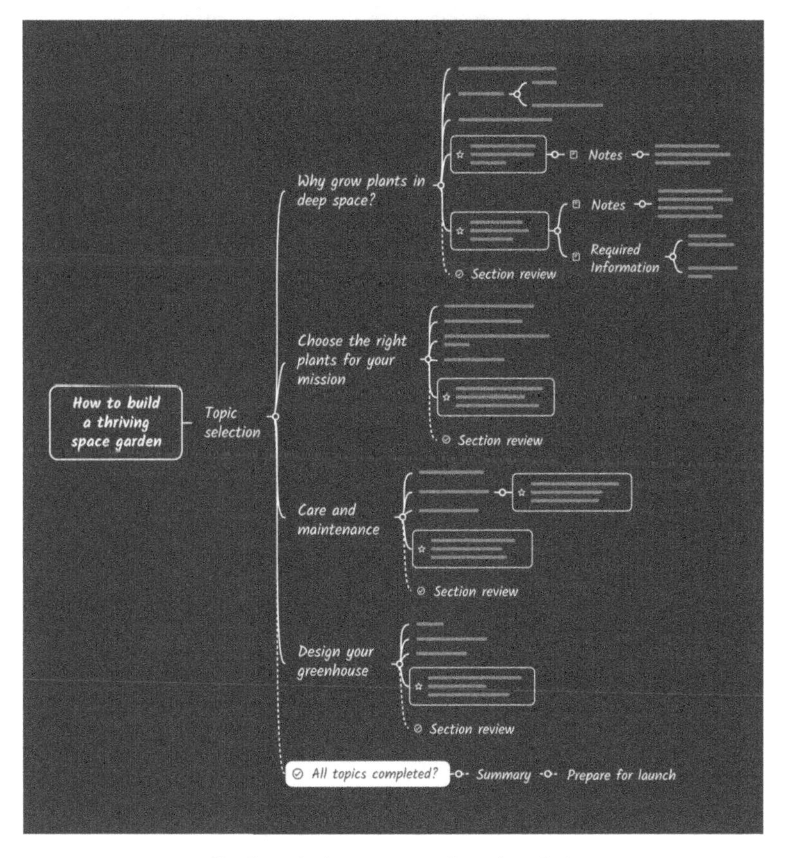

Check out Tools & Resources for a closer look.

As promised, here's a peek at the text outline equivalent. Which one would you prefer as your content map?

How to build a thriving space garden

- **Why grow plants in deep space?**
  - How long will food supplies last?
  - Improving health
    - Medicine
    - Psychological well-being
  - How much oxygen can we bring?
  - Interview: A funny thing happened on the way to the space station. (interstitial)
    - Notes
      - Share first-hand story from a stranded astronaut whose improvised garden helped carry them through.
  - Activity: Should you grow plants on this mission? (video scenario)
    - Notes
      - Present video interviews where astronauts discuss upcoming missions. For each scenario, users assess whether space gardens would be viable, helpful, and worth the investment.
    - Required information
      - Relationship between mission requirements and constraints.
      - Benefits of a space garden for humans.
  - Section review
- **Choose the right plants for your mission**
  - What will people need most?
  - What resources can you spare?
  - Are genetically-modified plants more resilient?
  - Fruits vs. vegetables
  - Activity: Select plants by mission requirements, environment, and resources. (matching exercise)
  - Section review
- **Care and maintenance**
  - Watering in zero gravity
  - Lighting without the sun
    - Activity: Configure UV intensity and timing based on species. (interactive simulation)
  - Preventing microbe growth
  - Activity: Assess plant health and diagnose common issues. (multiple choice w/images)
  - Section review
- **Design your greenhouse**
  - Substrates
  - Irrigation and hydroponics
  - Lights and heaters
  - Activity: Configure your greenhouse based on conditions. (drag and drop components)
  - Section review
  - All topics completed?
  - Summary
- **Next steps**
  - Prepare for launch!

Oh, and NASA has some great articles on space gardening if you need tips.

Meet Gagné

Gagné's Nine Events of Instruction describe ways to make a learning experience more effective (Gagné, 1985). It's a tried and tested model that dates back to 1965 when Robert Gagné published *Conditions of Learning*. Today it's still used to improve instruction at all levels, whether you're designing lesson plans for kids or training adult employees.

Why are we talking about it here? Well, Gagné's helps ensure you have all the ingredients for stickier content. You can use it as a checklist during outlining, storyboarding, and prototyping.

What are the nine events? So glad you asked!

1. Gain attention

Fact: it's widely acknowledged that Martian fudge is not only the tastiest in the galaxy but—when properly chewed—it makes you transparent for three-and-a-half hours. Did that pique your interest? I freaking hope so. If chocolate and aliens don't grab you, you're clearly beyond help.

Gagné's first event states that before learners can absorb anything, you need to get their attention. You might use humor or eye-catching activities, write an unusual introduction, lead with a question, or arouse curiosity through problem-solving.

Surprise, startle, provoke, or intrigue; do something to get your learners' minds in the game.

2. Inform learners of the objectives

After capturing attention, instructors should state the objectives of a learning experience. Basically: "Tell me what I'll be learning and what I should be able to do by the end."

This doesn't need to be complicated. You could share topics before diving into lessons or list the criteria for success. Knowing objectives prepares your learners for the journey ahead. It shows them what to expect and justifies their effort.

3. Stimulate recall of prior learning

New knowledge builds on what we learned previously. You can't learn to write before you understand spelling, and you can't spell squat until you know about letters. Connecting new material with past experience makes it easier to absorb.

To pull this off, add earlier topics to your questions and references (remember *activation* from Merrill's Principles of Instruction?). Analogies are also useful for linking new and existing information. The human brain weighs as much as a plump Chihuahua! Walrus whiskers feel like uncooked spaghetti! Call out something learners can relate to, then use it to teach something new.

This method has all kinds of neuroscience behind it, but in short, you're making new connections faster by accessing old ones. Walking a well-trodden path is easier than bushwhacking through the wilderness.

4. Present material

With learners all primed and ready, it's time to present your content.

During this event, instructors share new material through whatever means and methods work best. Pair topics with examples, organize and chunk your content, and use various media types for those who absorb information differently. Apply all the instructional techniques we've learned to show off what you're teaching.

5. Provide learning guidance

Like any good instructor, you should help learners find the best path through your content and support them when they get stuck. In eLearning, you might have them follow examples, offer hints, and give them tips to retain information. Real-world case studies are useful, as is corrective feedback.

The *Designing for memory* chapter may help here.

6. Elicit performance

This one's about proof and practice. Have learners demonstrate new knowledge through scenarios, quizzes, presentations, and probing, essay-style questions.

Think you understand the material? Show me!

7. Provide feedback

Feedback is a vital part of the learning experience. Without it, learners don't know what they've done right or wrong, making it tricky to course-correct.

Adding feedback to your activities is a great start. Ask questions, dissect the answers, and have people rework what they missed. In more nuanced scenarios, you can help learners see why one approach might be preferable to another.

Self-evaluation is another tactic. By telling people what they should have learned at the end of a lesson, they can decide if they need to review.

As you write feedback, try to balance critique with encouragement. You want to nudge learners in the right direction without crushing their spirits.

8. Assess performance

Assessments help measure whether a learning experience has the intended impact. You might test learners before and after, sprinkle quizzes throughout, or analyze LMS data. As you update your training, you can compare the results over time.

9. Enhance retention and transfer (to the real world)

As part of Gagné's 9th event, you'll help students carry their knowledge from the learning environment to the real world. You'll prepare learners for this transition through practice, reference aids, and imagination.

You might ask learners to describe how they'll apply the material moving forward. When you wake up tomorrow and sit at your desk, what will you do differently based on this training?

Quick-reference job aids are also helpful, giving learners handy, bulleted summaries they can skim until the material becomes second nature. This might include posted recipes from our sandwich shop or a list of questions for customer support.

We've already talked about using scenarios for introducing real-world context. The more realistic your stories and practice exercises, the easier it is to transpose skills later.

Applying Gagné to your outline

Now that you've got Gagné's events handy, you can start putting them to work. If you review your outline and the event list side-by-side, does it spark new ideas? Look for places you might add feedback, preview what's coming, or help learners practice. Your outline is a great place to see a broad view of the whole.

Chapter takeaways

Outlines help you organize material from SMEs, highlight gaps to fill, and see where you can reinforce learning with exercises. Once you structure your content, you can move to a higher level of fidelity and work out the details.

That's right. It's almost time to storyboard, people!

- *Outlines* help structure your eLearning experience. Use them to organize your thinking and plan the flow of your content.
- Before starting your outline, work with stakeholders to define project goals and gather a threshold of material.
- *Traditional* learning presents learners with information, then reinforces it with quizzes or exercises. *Scenario-based* learning places people in realistic situations, asks them to make decisions, and reveals insights with each outcome.
- eLearning varies from conventional materials in that people can navigate non-linearly. So, in addition to specifying a content sequence and hierarchy, eLearning outlines should describe possible user paths.
- You can format your outline as a simple text hierarchy or as a visual map. While each approach has pros and

cons, the visual outline has advantages for non-linear, digital storytelling. Plus, it feels more creative!

- If you didn't already pick a digital tool, choose software that aligns with your outline format. Easy grouping and reshuffling of text is a must.
- *Gagné's Nine Events of Instruction* define characteristics of effective learning. Using them as a checklist helps ensure your outline has all the right ingredients.

12

LEVEL OF FIDELITY

Don't get too comfy. This chapter should be quick, but it may help with your planning.

In design terms, an *artifact* is any document or deliverable produced from a stage of work. Outlines, storyboards, wireframes, design layouts, and prototypes are all artifacts.

Before working on any artifact, you'll want to determine its underlying purpose. This helps you decide what to include and how detailed it should be. These factors make up an artifact's *level of fidelity*.

Since no project has unlimited time, budget, or resources, you'll want to choose the right level of fidelity for each step.

Low-fidelity vs. high-fidelity

Depending on your goal, various levels of fidelity are possible.

Low-fidelity steps should feel sketch-like, rough, and unpolished. They require less time and can be iterated (and obliterated) quickly. Add ideas. Re-order. Slash and burn. You might use placeholder content, and nothing should feel overly precious if you choose to reshape or re-organize. Generally, you'll work in low-fidelity when sharing early concepts, brainstorming, or moving ideas around to see what works. Sketches and outlines tend to be low-fidelity because they're easily created and tweaked.

High-fidelity artifacts are polished and more closely resemble a finished product. Questions have been resolved, details fleshed out, and work approaches reality. Design layouts are often rendered in high fidelity since they're meant to simulate what learners will see, and to get final feedback before development.

And yes: *medium-fidelity* is also a thing.

Increase fidelity over time

As details grow clearer, you'll want to progress from low-fidelity to high-fidelity work.

For instance, you might share evolving documents with your stakeholders. You could start by drafting topics and order (low-fidelity), then write on-screen text (medium-fidelity), then list final video assets and details for each click (high-fidelity).

It's worth noting that some artifacts, by nature, are higher-fidelity than others. When designing eLearning, you may move from sketches to wireframes to polished layouts.

Match the level of fidelity to the goal

For each deliverable, you'll want to match the fidelity to the purpose. Consider why you're creating something and what you need to accomplish. What feedback do you need from reviewers? Where do you want them to focus? Are any details irrelevant right now?

Suppose it's early in a project, and you're deciding which scenarios might resonate with learners. Confident in your choices, you invest time and budget in a high-fidelity storyboard, complete with illustrations and dialogue, before discussing narratives with your client. Three minutes into your presentation, the client says they've tried your approach before, and employees found it deeply unhelpful. Whoops!

What you needed from that exercise was a list of viable, stakeholder-endorsed scenarios to develop further. Instead, you spent all your time polishing a single line of thought. Your artifact didn't match the goal, and you may have just blown the budget futzing with illustrations when a ten-minute sketching session (or even a conversation) would've directed you elsewhere. Not ideal.

Sure, a calculated risk can pay off. High-fidelity work requires less imagination from the audience and might be compelling, but it also burns time and feels less open to input and revision. And if your client dislikes it, they may ask why you invested so much effort before consulting them.

On the other hand, sharing low-fidelity work can strengthen your relationships with clients, SMEs, and stakeholders. It invites feedback quickly and, because you're showing work in progress, helps them feel involved. The more they participate (within reason), the fewer surprises at the end.

The nature of unpolished work also directs attention. If all the people in your design appear as loose gesture drawings, you're clearly not looking for comments on photography in this review!

It's 100% acceptable to have some parts of your work rendered in detail while others remain sketchy. For example, if your goal is to collect feedback on navigation, you might invest in high-fidelity designs of your user interface. Conversely, on-screen text and imagery may be irrelevant here, and you could just use content placeholders.

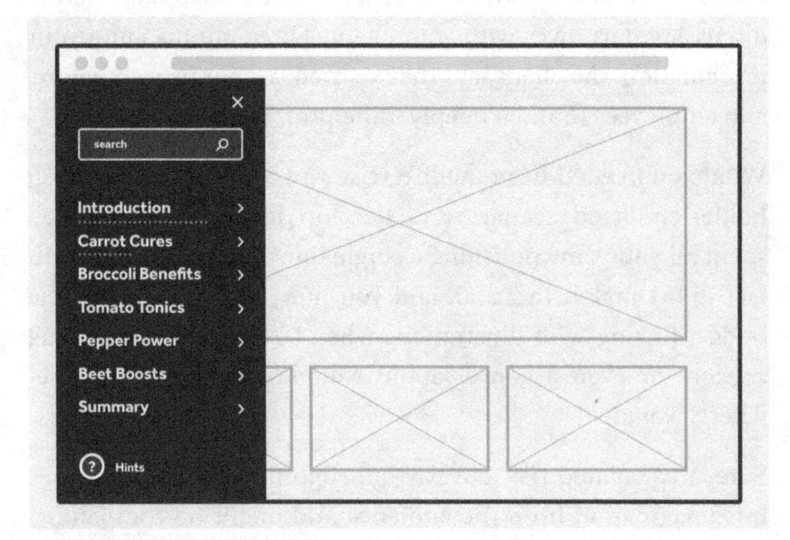

The critical point is that for each artifact—at each project phase—you're deliberately choosing a goal and level of fidelity in advance. You'll consider which features are worth investment and which to reserve for later.

How can you sum this up?

- I am creating this artifact with the goal of ________.
- To do this effectively, I will invest the most effort in ________.
- It may be less important to spend time on ________.

Here's an example for clarity:

- I am creating this artifact with the goal of *getting client approval on how the navigation looks and functions.*

- To do this effectively, I will invest the most effort in *menu titles, annotations, and UI design.*
- It may be less important to spend time on *the layout or text for the surrounding experience.*

Also, if your answer to the first two blanks is "everything," your tasks have been poorly distributed. Time to reprioritize!

Prep your stakeholders before you present

Again, your goal is to create a progression in the project workflow. As you review work and gain insights, you can increase the level of fidelity. Each iteration adds detail, requires time, and makes backtracking more difficult.

There are (at least) two great reasons to make your client aware of the level of fidelity before a presentation.

1. So they know what to expect and can give appropriate feedback.
2. So they know their approval means you'll be moving forward to the next level of fidelity. Client approvals have consequences and make your work more solid. Things can't stay fluid forever.

Remove distractions to the current task

When working with others, editing documents or templates along the way can be helpful. For example, you might omit unused fields during low-fidelity tasks to keep things focused. If you're working with a client to discuss scenarios, having empty spots for image or video files may be distracting. You won't review these details yet, so why show them?

Chapter takeaways

What's the moral here? Considering the level of fidelity before each task can direct your efforts and set client expectations. Stay loose and quick while sketching broad strokes; fine-tune as your answers sharpen. And make sure your audience sees only the details you need them to review.

Short chapter. Told you!

- *Artifacts* include any document or deliverable produced from a stage of work.
- You can create artifacts at various levels of detail and accuracy, also called *levels of fidelity*.
- *Low-fidelity* artifacts require less detail and investment, while *high-fidelity* artifacts can be time-intensive and resemble a finished product.
- Before working on any artifact, define its purpose and match your level of fidelity accordingly.
- An artifact's level of fidelity should increase over time as you make decisions and answers grow clearer.
- When working alongside clients and stakeholders, keep them involved, set expectations, and keep them focused on what you want them to review. Choosing an appropriate level of fidelity at each step can make working relationships smoother.

13

EVOLVE THE OUTLINE INTO A STORYBOARD

You've created your outline. Now let's put some flesh on that skeleton and turn it into a full-fledged *storyboard*! First, we'll decide on content and format. Then we'll add visual assets, on-screen text, voice-over, and interactivity, so everything's ready to build.

What's an eLearning storyboard?

Chances are, you've seen storyboards from movies, animation, or game design. In eLearning, storyboards serve much the same purpose. They show the narrative of a module or course, highlight when, where, and how you'll use visual assets, and provide functional notes for the developer. If an outline creates a map of your content, the storyboard shows what happens at each stop.

Create a storyboard document

You need a way to describe what a learner sees, reads, hears, or does at each point, arranged in a logical sequence. But eLearning comes in many forms, and you should lay out a storyboard based on what's most important. A storyboard for talking-head videos, for example, will have different fields than branched scenarios, quizzes, interstitials, and games.

Each snapshot from your storyboard is called a *frame*, and each state from eLearning is often called a *slide*. We'll use them interchangeably here since one storyboard frame usually corresponds to a single eLearning slide. Similarly, the words *field* and *cell* both refer to a single piece of storyboard data. For example, "On-screen text" is a common field. How you arrange these fields varies based on your needs.

Simple eLearning may rely on visuals with little user interaction—say, a video series where a user clicks "Next" to move through your content. In this case, storyboards might need only thumbnail sketches and short, descriptive blurbs.

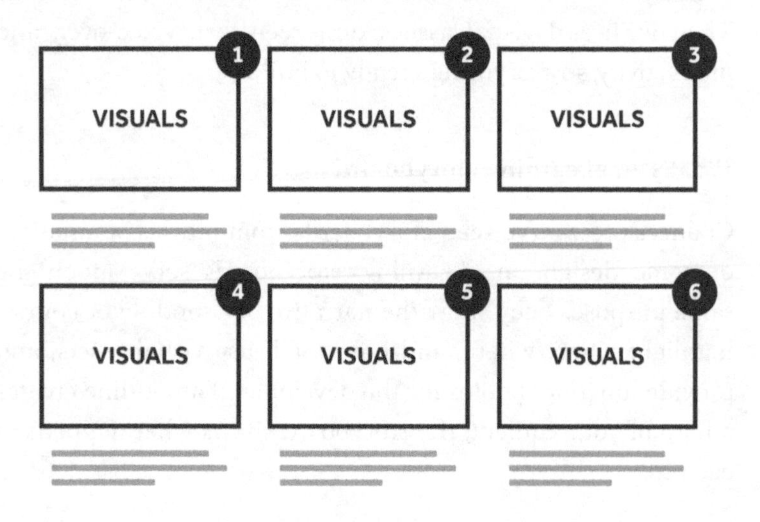

Others may combine visuals with on-screen text, adding transitions or interactivity that require developer notes. In this case, a frame might need more real estate—and more fields—to tell the story.

VISUALS
1
ON-SCREEN TEXT
DEV NOTES

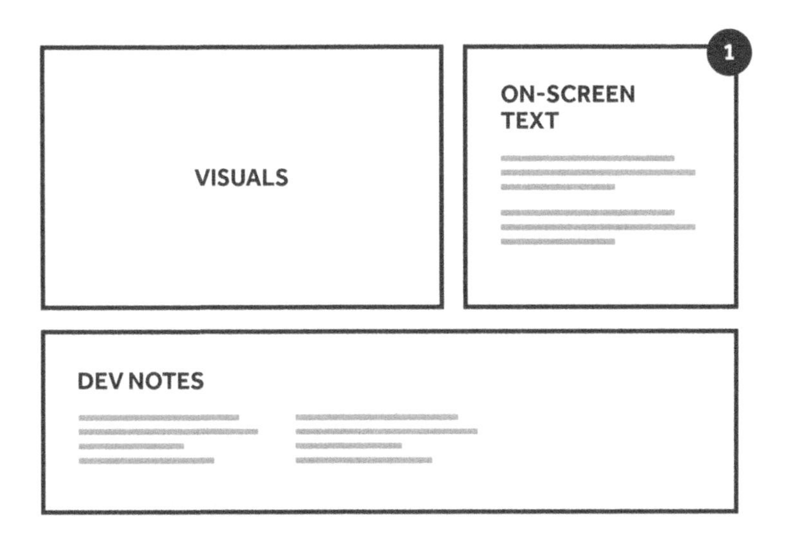

As complexity increases, so does your storyboard layout. You can design more detailed eLearning storyboards as tables, grouped into slides, with cells for each piece of data.

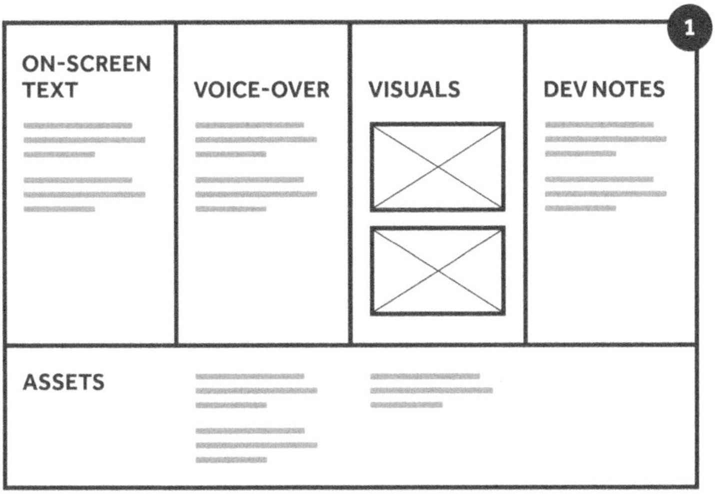

1
ON-SCREEN TEXT
VOICE-OVER
VISUALS
DEV NOTES
ASSETS

Your intended audience is also worth considering. Are you writing storyboards for your own reference? Then sketchy, low-fidelity docs may be fine. However, if you plan to share them with eLearning developers or clients, you'll need more polish and explanation, especially if you won't present them in person.

In this chapter, we'll look at a fairly robust example. Once you're comfortable, you can shuffle things around and remove unnecessary fields. The process is highly adjustable.

What fields do you need?

Again, a storyboard is made of individual slides representing moments in time. Fields in your document describe what happens on each slide, and your outline helps connect the dots.

Hopefully, you've chosen your tools from the early days of the *Get organized* chapter, but if not: grab one app for tables and another for basic sketching (and yes, old-school pencil and paper are okay). You don't need to be an illustrator; this is just a means of capturing visual ideas quickly. Thumbnail sketches are entirely optional.

Using your outline and software of choice, create a table with fields you know you'll need. My recommendations are as follows:

Basic storyboard fields

- Module, section, or chapter titles
- Slide number
- On-screen text
- Voice-over (audio narration script)
- Visuals (notes, sketches, or placeholder images)

- Dev notes (guidance for UI)

Detailed storyboards may add the following

- Assets (names or links for media files)
- Layout, animation, or interaction notes

File tracking is also critical. A good naming convention makes it easy to distinguish clients, projects, and rounds of revision. For example, you might name files with the following formula:

Formula: *[client acronym]-[project name]-[deliverable]-[review round #]*

Example: *aub-spaceplants-storyboard-r01.doc*

And yes, as long as your team can find what they're looking for, feel free to abbreviate.

Storyboard workflow

Once you've created a storyboard document, update the round number in your file name. If this is a fresh storyboard, this would be client round 1. You'll increment that number with each revision. For example:

- *aub-spaceplants-storyboard-r01.doc*
- *aub-spaceplants-storyboard-r02.doc*
- etc.

Review your outline and create storyboard slides (or table cells) for each major beat. Keep in mind that some beats may require multiple slides while others need only one. You can add more later, but for now you have space to write.

0.0 - Introduction			
On-screen text	Voice-over	Visuals (notes, sketches, or images)	Dev notes
Assets			

And so it begins, with a single slide.

Assigning slide numbers can be tricky, but you need a way to reference specific blocks in your storyboard for review, feedback, and UI links.

One method is to join sections and subsections in your naming. For example, the first slide in your first section would be slide 1.1, the second slide 1.2, and so on. A popup launched from the latter slide would be slide 1.2.1. You see the challenge—the more layers an experience has, the more depth you'll need. For obvious reasons, slide numbers can get confusing if you don't plan them early.

You'll also notice it's harder to re-sequence now that you're schlepping fields with each slide. That's why the outline was valuable. It's easier to move things around when you're baggage-free!

For each frame, start writing on-screen and UI text. You won't know every detail in this first pass—just draft a cohesive narrative and transcribe the flow from your outline. You can polish your language later.

0.0 - Introduction	
On-screen text	**Voice-over**
Title How to build a thriving space garden **Description** Due to the lack of gravity and sunlight, growing plants in space requires careful planning. However, gardens provide numerous benefits, such as fresh food, extra oxygen, and improved psychological well-being. Let's look at ways to design and maintain a garden for your next mission! **Button** Get started *[Links to 1.0 - Topic selection]*	

Add comments for each interaction. Where do buttons take you? Which choices are correct, incorrect, encouraged, or require adjustment? If UI launches a popup or links elsewhere, note that in brackets.

Let's look at ways to design and maintain a garden for your next mission!

Button
Get started
[Links to 1.0 - Topic selection]

UI link syntax

We'll look at more robust examples, but make sure this syntax makes sense.

Once you have a narrative blocked out, it's time to suggest content in the *Visuals* field. You can use this space in a few different ways.

Early in the project, you'll mull over ideas for each slide. Your visuals will be unknown, and this field can house notes for selecting, shooting, or creating assets (e.g., "Find an image of a person doing X"). You might suggest layouts with thumbnail sketches and make updates as you go. Regardless, the goal is to help others quickly picture your content.

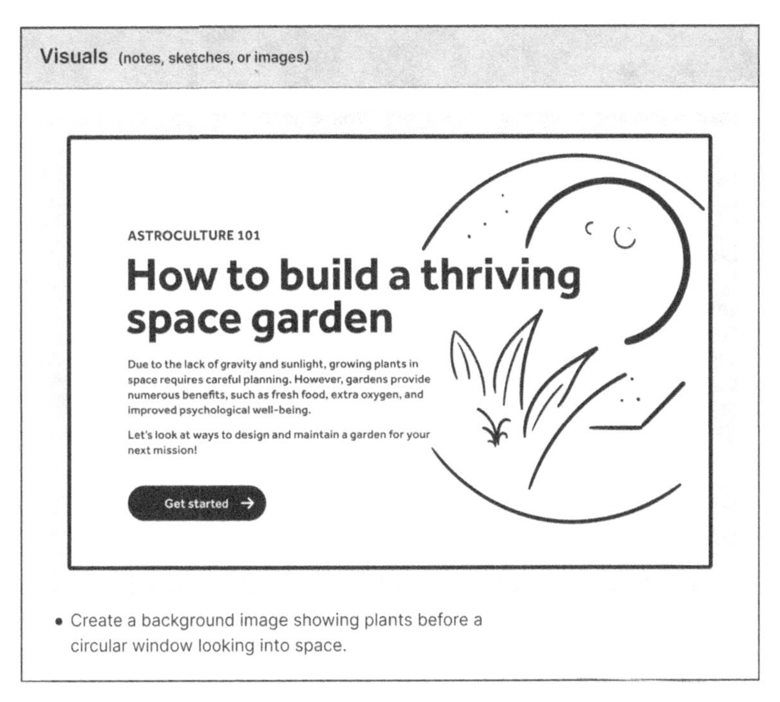

The Visuals field is flexible. You don't need to get this detailed.

If your eLearning includes voice-over, now's a good time to start drafting a script in the *Voice-over* column. You'll decide what on-screen text translates naturally into speech and what you should rephrase for conversation. Omit this column if you're not using voice-over (or if it always matches the on-screen text).

0.0 - Introduction	
On-screen text	**Voice-over**
Title How to build a thriving space garden **Description** Due to the lack of gravity and sunlight, growing plants in space requires careful planning. However, gardens provide numerous benefits, such as fresh food, extra oxygen, and improved psychological well-being.[1] Let's look at ways to design and maintain a garden for your next mission![2] **Button** Get started *[Links to 1.0 - Topic selection]*	Welcome to the Astroculture 101 course on building a thriving space garden! Due to the lack of gravity and sunlight, growing plants in space requires careful planning. But gardens can provide benefits for astronauts, like fresh food, extra oxygen, and improved well-being.[1] These lessons will help you design and maintain a garden for your next mission, so let's get started![2]

Remember, you may need to write alternate text for screen readers or audio-only experiences regardless of your eLearning format. Storyboarding is a perfect time to start thinking about accessibility.

After adding on-screen text, visuals, and audio, you can start drafting animation notes. For example, you might use matching superscript numbers to sync on-screen text with voice-over. Simple. Elegant. No writeup required.

Use the *Dev notes* column to elaborate—for example, to record an exact style of fades, flips, or translations. Add notes on interactivity that don't fit elsewhere.

Dev notes

Layout

- This slide uses the **Introduction** template.

Animation

- Fade in each text block to coincide with the voice-over.

Interactions

- Click the **Get started** button to progress to the **Topic selection** slide.

Throughout your project, your storyboards grow more detailed. You move from image wish lists to production file paths. Scripts are revised and approved by stakeholders. The level of fidelity evolves. Just remember to prepare your clients before presenting your storyboards. Don't let them misinterpret temporary text or visuals as final. You'd be shocked by how often this happens.

As you gather your media, list them in the *Assets* field for your eLearning developer.

Assets
Images • space-garden-intro-bg.png **Audio** • See the **0.0 Introduction** folder for .wav files.

A quick example

Here's a sequence based on our outline to show you how slides connect. I've restructured cells for easy reading, but you'll find a full view (and template) in the *Tools & Resources* section.

We start with an introduction, stating our course objective and what learners should be able to do by the end: design and maintain a garden for their next mission into space!

In our storyboard, we've added on-screen text, voice-over, recommended visuals, and notes for layout, animation, and interactivity. This slide has a single **Get started** button that links to the **Topic selection** screen from our outline.

0.0 - Introduction	
On-screen text	**Voice-over**
Title How to build a thriving space garden **Description** Due to the lack of gravity and sunlight, growing plants in space requires careful planning. However, gardens provide numerous benefits, such as fresh food, extra oxygen, and improved psychological well-being.[1] Let's look at ways to design and maintain a garden for your next mission![2] **Button** Get started *[Links to 1.0 - Topic selection]*	Welcome to the Astroculture 101 course on building a thriving space garden! Due to the lack of gravity and sunlight, growing plants in space requires careful planning. But gardens can provide benefits for astronauts, like fresh food, extra oxygen, and improved well-being.[1] These lessons will help you design and maintain a garden for your next mission, so let's get started![2]
Assets	**Dev notes**
Images • space-garden-intro-bg.png **Audio** • See the **0.0 Introduction** folder for .wav files.	**Layout** • This slide uses the **Introduction** template. **Animation** • Fade in each text block to coincide with the voice-over. **Interactions** • Click the **Get started** button to progress to the **Topic selection** slide.

Introduction fields

Your visuals don't need to be as detailed—these are essentially low-fidelity wireframes that imply layout. But as you can see, sketching makes content easier to explain and requires less imagination.

Introduction visuals

As users progress to the **Topic selection** slide, they can choose a starting point. We know why this is helpful for learning—because adults are self-directed, says Knowles!

Here, you can also see the syntax for multiple links.

1.0 - Topic selection	
On-screen text	**Voice-over**
Header Where would you like to begin? **Button list** • Why grow plants in deep space? *[Links to 2.1 - Why grow plants in deep space]* • Choose plants for your mission *[Links to 2.2 - Choose the right plants for your mission]* • Care and maintenance *[Links to 2.3 - Care and maintenance]* • Design your greenhouse *[Links to 2.4 - Design your greenhouse]*	Where would you like to begin? We can look at why astronauts might want to grow plants in deep space, how to select the right plants for your mission, care and maintenance, or greenhouse design!
Assets	**Dev notes**
Icons • icon-selection-arrow-01.png • icon-home-01.png **Audio** • See the **1.0 - Topic selection** folder for .wav files.	**Layout** • This slide uses the **Topic selection** template. **Animation** • Fade in each option to coincide with the voice-over. **Interactions** • Click a topic to progress to the first slide in that branch.

Topic selection fields

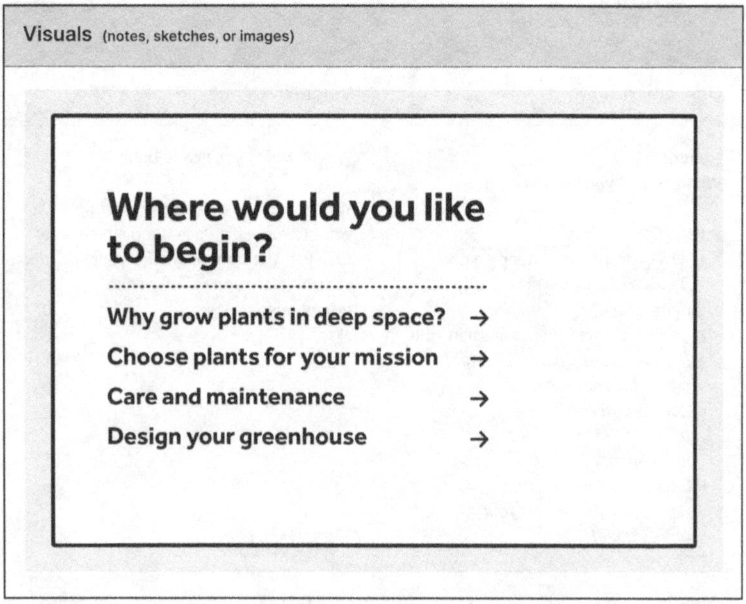

Visuals (notes, sketches, or images)

Where would you like to begin?

Why grow plants in deep space? →
Choose plants for your mission →
Care and maintenance →
Design your greenhouse →

Topic selection visuals

Flash forward to an activity for assessing plant health (also from our outline). Our plants aren't faring so well in this scenario, and we must decide what to do. This slide shows how to annotate options—whether they're right, wrong, desirable, or less so.

Our slide numbers are beginning to deepen. Look at this example and see if the digits track for you.

2.3.4 - Care and maintenance | Assessing plant health	
On-screen text	**Voice-over**
Header Well, this doesn't look good. **Description** Many of the leaves are turning yellow with brown spots. You wonder if it's a watering issue or a nutrient deficiency due to the unique growing conditions in space. What do you do?[1] **Option list** A. Plug the drainage holes so your plants get more water. *[**INCORRECT**, links to 2.3.4.1 - Result]* B. Add fresh fertilizer daily until the leaves return to normal. *[**INCORRECT**, links to 2.3.4.2 - Result]* C. Prune the discolored tips and make sure the soil is watered evenly. *[**CORRECT**, links to 2.3.4.3 - Result]*	Your team has carefully selected species and built systems to provide nutrients and water. However, you start to notice some issues with your plants. Many of the leaves are turning yellow with brown spots. You wonder if it's a watering issue or a nutrient deficiency due to the unique growing conditions in space. What do you do?[1]
Assets	**Dev notes**
Images • plant-health-wilted.png **Audio** • See the **2.3.4 - Care and maintenance** folder for .wav files.	**Layout** • This slide uses the **Multiple choice** template. • Breadcrumb navigation appears at the bottom. **Animation** • Fade in each option to coincide with the voice-over. **Interactions** • Click an option to see the result of each choice.

Activity fields

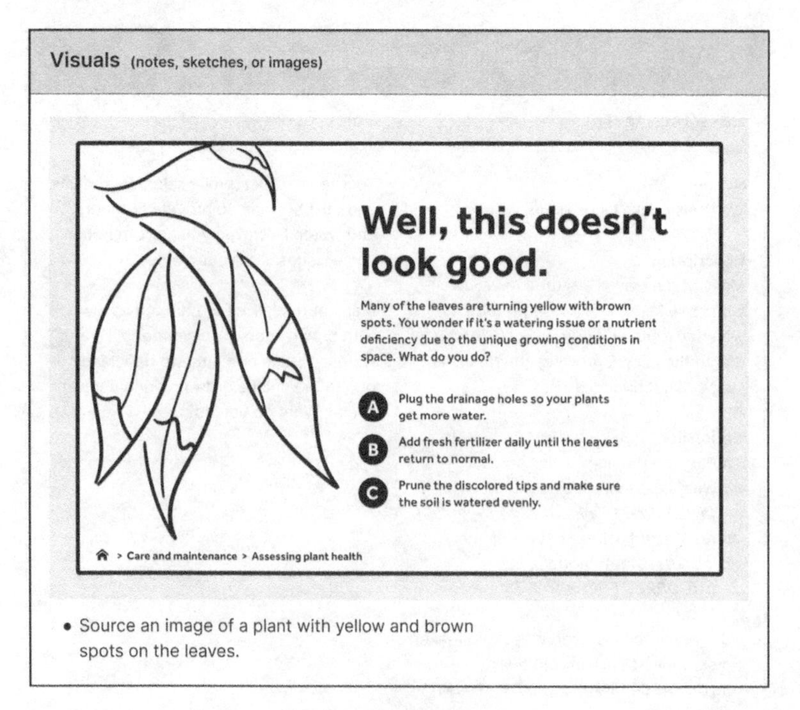

Activity visuals

Finally, we arrive at a result. Unfortunately, the health of your space garden declined because you overfertilized, but you've gained valuable feedback on ways to improve. Maybe you'll make a better choice next time.

<table>
<tr><th colspan="2">2.3.4.2 - Care and maintenance | Assessing plant health | Result</th></tr>
<tr><th>On-screen text</th><th>Voice-over</th></tr>
<tr><td>Header
Houston, we have a problem.

Result description
Unfortunately, the leaves continue to wilt and you lose 40% of your plants over the next few weeks.

Fertilizing plants every day can actually do more harm than good. It can lead to nutrient burn and damage to the plant's roots.

Button
Retry
[Links back to 2.3.4 - Assessing plant health]</td><td>Houston, we have a problem.

Unfortunately, the leaves continue to wilt and you lose 40% of your plants over the next few weeks.

Fertilizing plants every day can actually do more harm than good. It can lead to nutrient burn and damage to the plant's roots.

Want to try something else?</td></tr>
<tr><th>Assets</th><th>Dev notes</th></tr>
<tr><td>Video
• plant-health-overfertilizing-time-lapse.mov

Audio
• See the 2.3.4 - Care and maintenance folder for .wav files.</td><td>Layout
• This slide uses the Result and feedback template.

Animation
• Fade in each option to coincide with the voice-over.

Interactions
• Click the Retry button to return to the scenario screen.</td></tr>
</table>

Result fields

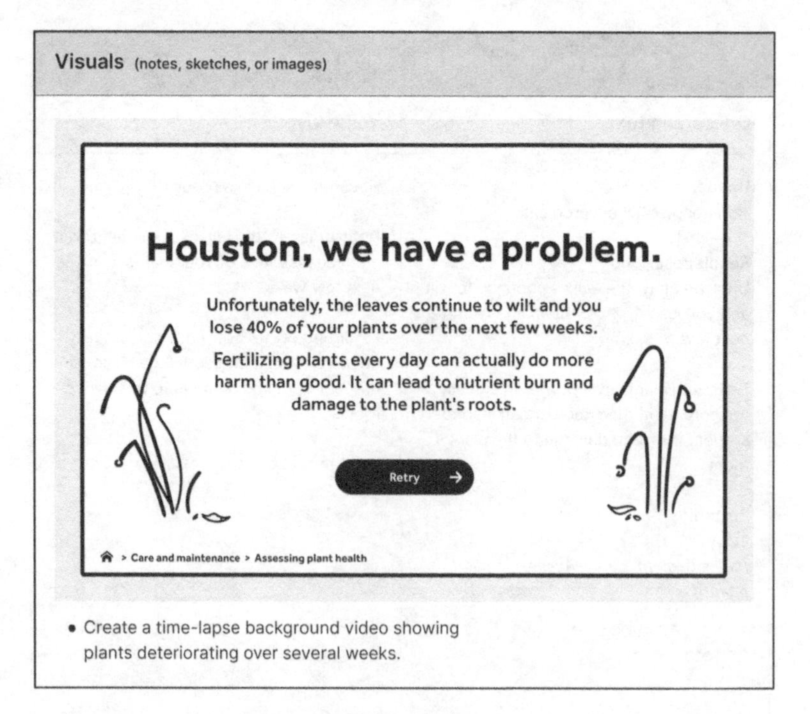

Result visuals

You won't populate the **Assets** column until exporting your final media, but these examples should help you prepare for development.

Remember, good media keeps learners immersed in your narrative. Sure, you could write all your outcomes as lonely text, or you might get more creative. Who *wouldn't* want to watch time-lapse video of their space plant?

And since media assets are vital (and volatile) ingredients, we'll explore them next.

Sourcing your media

Your visual assets affect everything from schedules through development. Get them right, and your project goes much smoother. Underestimate them, and your budget's on fire.

Two questions help guide these efforts. First, what media types will you use—video, images, illustrations, or none of the above? Second, do you have the means to create custom assets, or will you find and license stock media? There's a huge difference between building a text-only experience and shooting video for every scenario.

Below are some prompts to discuss with your team. I've listed them here because they relate to the storyboard phase, but you'll want to raise these as early as possible.

- Will this project require video, audio, illustrations, or still images?
- Will you or your team produce assets from scratch based on any scenario you dream up? If so, you'll account for casting, script writing, custom photography, or shooting video.
- Are you limited to finding and buying pre-made stock images? In this case, you'll hunt down assets that match your scenarios, then get client approval on your choices. While this can be less pricey than custom assets, the process can be shockingly time-consuming. Confirm your sources and budget for stock media, then add time for stakeholder reviews.
- Does your eLearning software have character libraries? If so, you might bypass some of the pitfalls of custom assets or stock searches. Libraries may include illustrated and poseable characters or photos of people

in wide-ranging scenarios. While the style and quality vary, these may be worth a look.

These questions should clarify which assets you need to find or produce. Once you have answers, you can plan accordingly.

If you think it's great: template!

Customizing a well-fitted storyboard document can be time-intensive. So, after you've molded one to your needs and tested it in your workflow, save that treasure and reuse, refine, reuse.

You can also find plenty of free templates to modify at will. There's a community of talented instructional designers out there, and everyone has their preferences.

Revisiting Gagné's Events of Instruction

Remember how we applied Gagné's Events of Instruction while outlining? Surprise! They're handy for storyboards as well. Use them to ensure your shiny new content is learning-ready. You might even add a column noting which Gagné's events relate to each exercise or story beat.

Chapter takeaways

Well-crafted storyboards are essential to the eLearning process. You'll use them to guide development, keep track of media, and get client approval on all your material. Some of the most important takeaways include:

- eLearning *storyboards* define the assets, content, and narrative for your learning experience.

- Storyboards contain *slides* representing moments in time. Each slide has *fields* with various kinds of information.
- There is no universal storyboard layout. Instructional designers format these docs based on preferences and the needs of each project. You can use the example from this chapter or find others from the ID community.
- Slides in your storyboard inherit content from your outline. Begin with minimal data, then increase detail as you go.
- Consider your visual assets in advance. Make sure you have the resources to find or create all the media you need.
- Once you've settled on a storyboard format, save it as a template for future projects.

14

VISUAL DESIGN: A CRASH COURSE

So far, you've built the meat and bones of your content—outlining, storyboarding, and planning. Now it's time to put some clothes on this body of work by applying visual design. But it's not just about dressing things up or making them look pretty (although that's certainly part of it). Good design builds on long-established pillars to make your work feel intentional and help viewers absorb information.

What the heck is good design?

Much like art, visual design has rules that should be learned before they're broken. Good design applies these principles to feel intentional to the viewer, user, or reader. It's the difference between something that looks balanced and organized, and a garbled, chaotic mess!

In the context of eLearning, good design makes it easier for learners to distinguish information. It saves mental energy for where it's needed rather than burning it on re-grouping or sorting content (remember cognitive load theory?).

Good visual design does some of the brain's work in advance.

Principles of good visual design

The exact number of design principles varies, mainly because people combine them or use different terms. But most seasoned designers will agree on *contrast*, *repetition*, *alignment*, and *proximity* as pillars of their practice. Other principles add nuance and generally extend the first four. For example, you might create *hierarchy* using contrast, alignment, or proximity. But we'll get to that.

Contrast

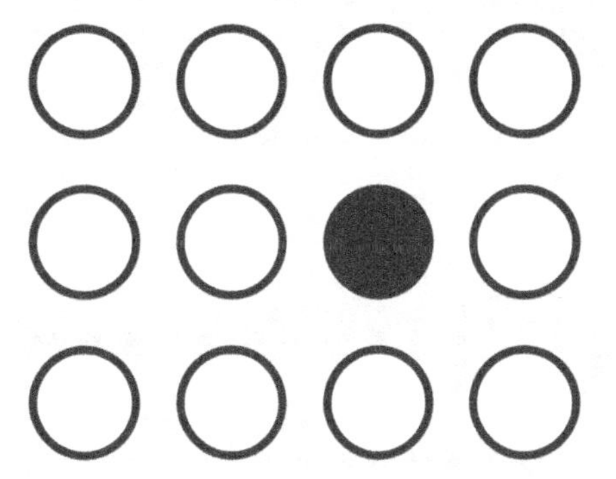

As you'd expect, *contrast* emphasizes the differences between visual elements in a composition. You can introduce it by varying color, size, shape, texture, and other properties to build visually dynamic layouts. One thing is unlike another, and that catches the eye.

Contrasting elements direct attention and lead the viewer to notable parts of the design. A hot pink circle in the middle of a black field screams: **"LOOK AT ME!!! I'M IMPORTANT! I STAND OUT!"** Obviously, your execution can be much more subtle.

You might set a big, heavy typeface above small, delicate letters or use unexpected combinations—like placing smooth textures beside rough ones—to create separation. In either case, you're steering the observer deliberately.

Two words of caution here. First, if you try to make everything stand out, nothing stands out. Second, ensure your contrast is strong enough to be clear. For example, if you scale one element to be larger than another, but they're still close-ish in size, the contrast might seem unintentional and your message may be lost. This introduces *conflict* rather than *contrast*, which is more distracting than helpful.

Repetition

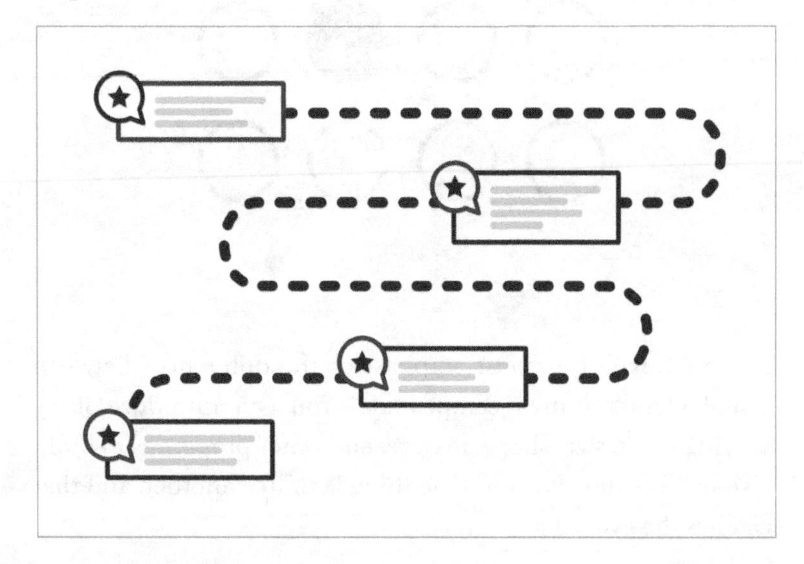

You guessed it! Repetition is the process of repeating visual elements to build consistency and structure. You can apply it to fonts, shapes, objects, icons, colors, images, and even words.

Repetition serves multiple purposes in design. You can use it to establish rhythm, show relationships, and build patterns. When you reuse design elements in a layout, you reinforce the idea that you deliberately planned this out.

Much like a chorus in music, repetition helps create expectation, even if your viewers don't consciously notice. They begin to understand: "When I see X, I can expect Y." For example:

- Whenever I see a blue circle with an icon, I know more information is available.
- One-pixel gray lines separate a quiz question from available answers.
- Featured article thumbnails always have the same dimensions and aspect ratio.
- Whenever I see headers with a specific font size and treatment, I expect the start of a new topic.
- Article paragraphs will always use a 16 px Georgia font.

Even this chapter is a basic example of repetition: chapter titles use one font treatment and subtitles another. You'll find this pattern over and over as you scroll down the page. Consciously or not, when you saw the Repetition header in a specific font weight and size, you (hopefully) knew this was the start of a new section and expected paragraphs below.

To think of it another way, picture the reverse. If every element in a layout was unique and there was no repetition, a composition would look chaotic. You'd constantly try to discern how pieces were related, and the experience would feel random. You'd exhaust yourself trying to organize information in your head because the designer (assuming there was one) handed you a mess.

Repetition is about balance. Designers should pay close attention to the color, size, and placement of repeated elements to ensure a composition feels cohesive. If something appears too often, it may become overwhelming or distracting. But if there isn't enough repetition, viewers may find no patterns in your design.

In instructional design, repetition can make information easier to absorb, enhance memory, and improve recognition.

Alignment

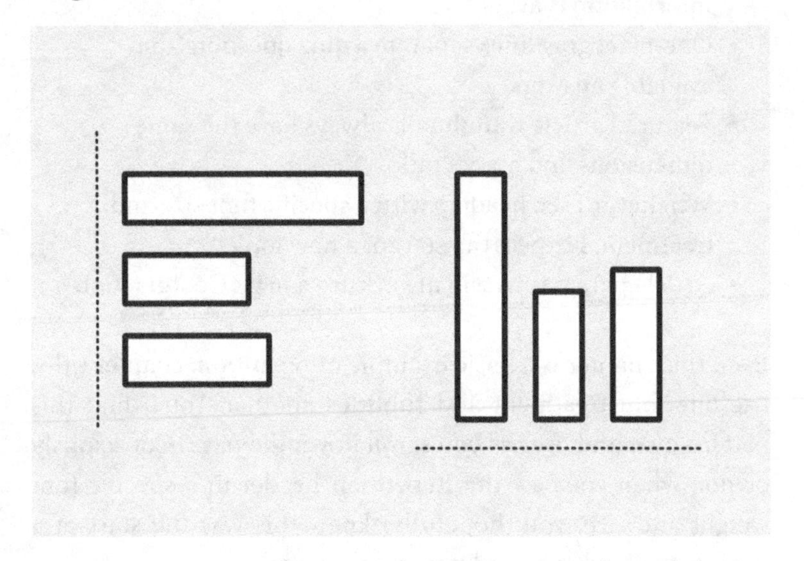

That's right! Don't be sloppy. Line things up. Good alignment creates a sense of hierarchy and structure. Lousy alignment will make your design feel disorganized.

We'll discuss your overall grid system later in this chapter, but the idea is to ensure nothing's unintentionally floating. Try to anchor elements by aligning some part of their form—especially elements with a relationship.

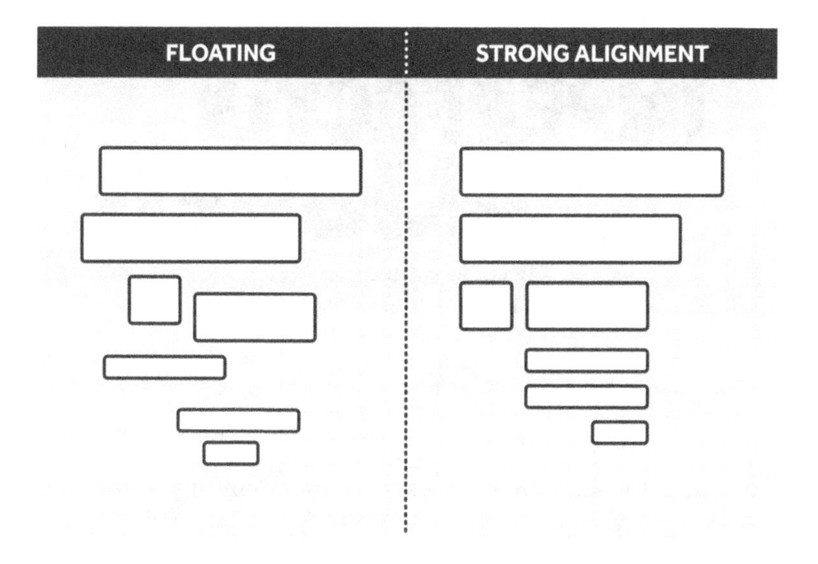

When you're first starting, focus on aligning outer edges, both horizontal and vertical. Make sure paragraphs align with headers, buttons, and interface elements where appropriate. Be meticulous here. Even a few misaligned pixels can draw attention and feel like a mistake.

As you get more comfortable, you can align more subtle attributes of shapes and curves. Shoving everything to one side can grow monotonous, so it's sometimes refreshing to shift your baseline.

IMPERFECTLY
ALIGNED

Because sometimes
the outer edges are boring.

You can use alignment to express hierarchy and grouping. Think of a paragraph followed by bullet points. What if one bullet wasn't aligned with the others? You'd instantly question if it was a mistake or if that bullet didn't belong with its siblings.

How do you ensure good alignment? Grids and guides from your design software are critical, but the most valuable tools are your eyeballs. Does anything seem like it's floating without (literal and figurative) justification? Can you strengthen relationships by aligning certain components?

Trust your gut. If something feels off to you, your learners will notice it as well.

Proximity

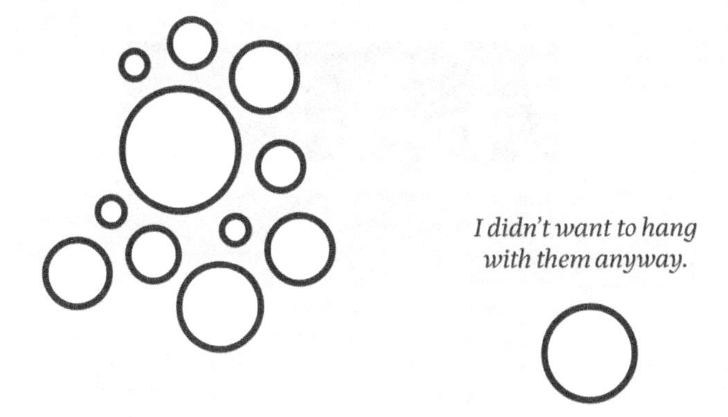

If you entered a room and saw people standing shoulder to shoulder with their arms entwined, you'd probably infer they had a relationship. They've breached personal bubbles and appear closer than strangers would usually be. But if you saw a person standing alone across the room, you wouldn't necessarily group them with the others.

That's the essence of proximity: physical closeness can imply a relationship between one thing and another. You can use it to group similar items with a unified message or, by increasing space, separate unrelated ideas.

Here's a simple example: when you visit a news site, a featured story might include a thumbnail image, a descriptive text blurb, and a read more button. These elements are positioned together because they all relate to the same feature and a single action.

Hierarchy

The main purpose of hierarchy is to create structure. What should the viewer pay attention to first, second, and third? What element is a sibling or child of another? You can establish hierarchy using all the preceding principles, but the intent is to guide viewers through information in a deliberate order.

This concept is easy to grasp because we constantly see it in text. Place a giant, bold header above smaller, delicate text blocks. Which draws your attention first? Which is the title, and what are the details?

I'M SUPER IMPORTANT. LOOK AT ME FIRST!

Hey, I'm pretty important, too.

I'm less flashy, but still noticeable.

Oh, don't mind me. I don't shout or draw immediate attention. You can use me without distraction, and readers will notice me last.

Size, order, and position are no-brainer methods for building hierarchy, but you can also play with weight and color. And when considering why this is relevant to you, think about what would happen if learners read your content out of order. What if they couldn't distinguish between quiz questions and answers, or between core learning material and a copyright blurb?

Emphasis

Emphasis is an easy one. Decide on the most important elements of your composition, then draw attention to where it should be. You can apply emphasis using many other design principles (especially contrast), but the idea is to do so with care. If everything is hot pink, the blaring color loses its ability to stand out. So, use contrasting colors, sizes, weights, spacing, etc. Just remember: emphasis is relative to the surroundings.

This term is most helpful as guidance for others. "I'd like you to give the button more emphasis" would convey your feedback without micromanaging. Designers can add emphasis in several ways.

Whitespace

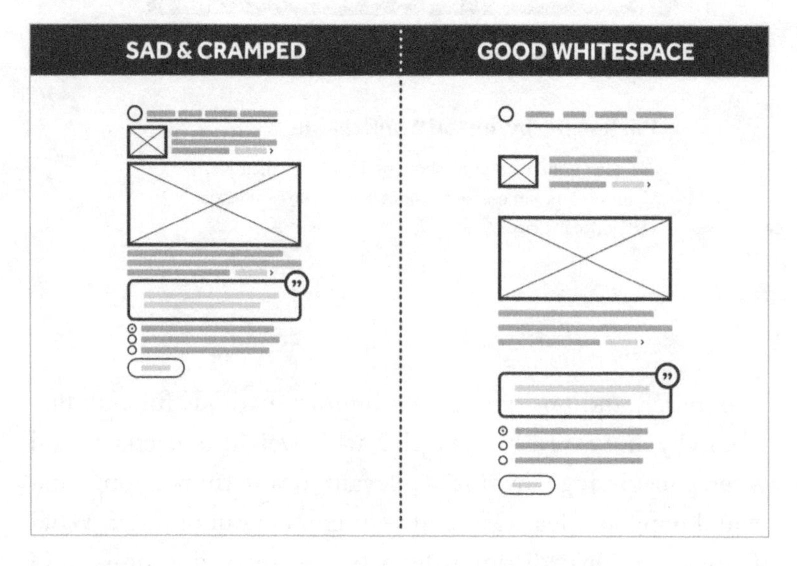

The age-old battle between designers and clients: "Can we please introduce some space to give the composition breathing room?" Well, that depends. Can you defend that precious territory with reasoning?

As the name implies, *whitespace* refers to the void between text, images, and other design elements. You can use it to direct attention, create balance, and emphasize specific aspects of a design. Good use of whitespace helps compositions look organized and professional. Most importantly, it makes information easier to distinguish.

By injecting space around an element or page layout, designers can separate blocks of information, which improves legibility. It's the difference between scanning aisles in a library and sifting through a hoarder's basement. In an organized setting, viewers can find what they're looking for and aren't overwhelmed by clutter.

It's worth noting that even though whitespace is literally blank, it still conveys relationships by way of proximity. Paragraphs with little space between them are likely continuations of the same topic; widely-spaced sections imply the start of something new.

Ultimately, whitespace helps add emphasis, create balance, and allow for more readable, distinguishable content. All of these reduce the effort for your learners.

Movement

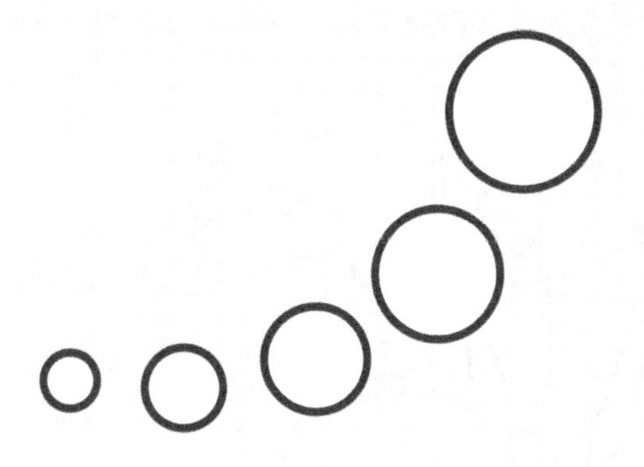

Wheeeeee!

This principle is more subtle. How will your composition guide the eye, and what path should it follow? Should it just move from top to bottom, or is there some *movement* you want to encourage?

There are many ways to influence movement through design. Using your audience's natural reading tendencies is a good starting point. Do your learners tend to skim left to right, right to left, or top to bottom when they read in their native tongue?

Case in point: if you're reading this in English, there's a good chance your eye starts at the top left and follows a Z shape as it travels. But what if you want to deliberately lead the eye in another direction, path, or order using purely visual cues?

The eye will naturally follow curves in your shapes and pinball between high-contrast elements. You can use this behavior to guide users and ensure their attention doesn't slide off a screen when you want to direct them elsewhere.

If I drew a giant curve on this page, your eye would naturally follow the shape. If I dropped three red dots from top left to bottom right, you'd follow them down a slanted, linear path.

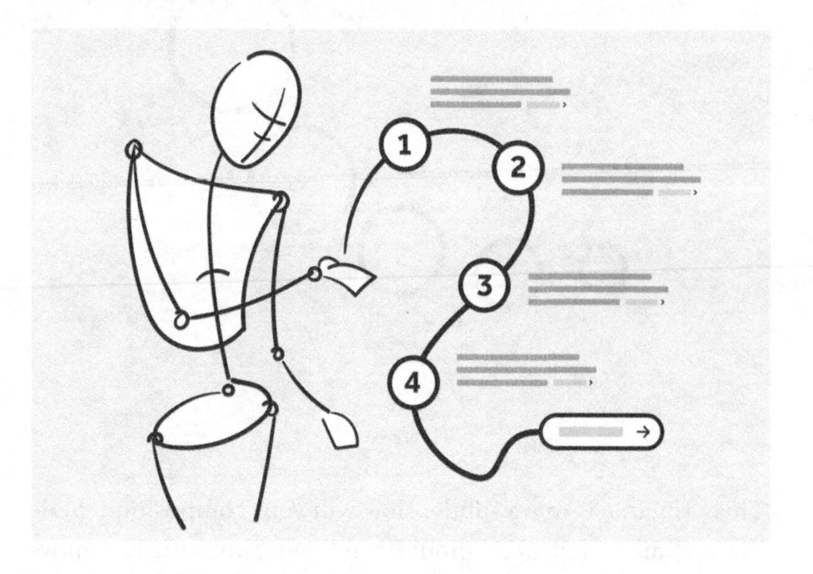

With complex layouts, this process grows more nuanced. But as you design each screen, note how your eye moves along. Does the path hit information in a logical order and move where you want it to go? Or does it slide off the page, making irrelevant stops along the way? When you're first starting, this process might seem abstract, but keep looking. You'll get a feel for it.

Ultimately, you want to create movement that guides learners along the path of least effort and highest logical progression. They should travel from beginning to end of your material, hitting all the major beats.

Comps

Time out! I need to define one thing before we continue.

Comps are design files where you'll apply visual design to your UI, content, and learning materials. Think of a comp as a single, high-fidelity design layout—a flat screenshot of your experience before it's built. It might show a site homepage, a screen for an eLearning experience, or an interface for a game leaderboard. The term is both a noun and a verb, so when you open a design file and start pixel-pushing, you're technically comping some comps! I know. Sounds weird. I didn't invent language, people.

Comps serve as polished, non-functioning visuals to review before building the final product. You'll create them in design software and output shareable media for others.

When collaborating with clients and stakeholders, you might produce batches of comps for approval before moving into development. They'll help you visualize content, work through UI, and supply devs with guidance and assets. For example, you might first comp a homepage and an internal page to present a concept before designing the entire site. If your client has feedback or dislikes what you've done, you only need to revise (or scrap) two comps rather than dozens.

Fun fact: the word comp comes from the word comprehensive, although people will sometimes claim it originates from composition or composite. You'll rarely find any of those words in a professional setting, but you'll hear comp all the time.

Grids

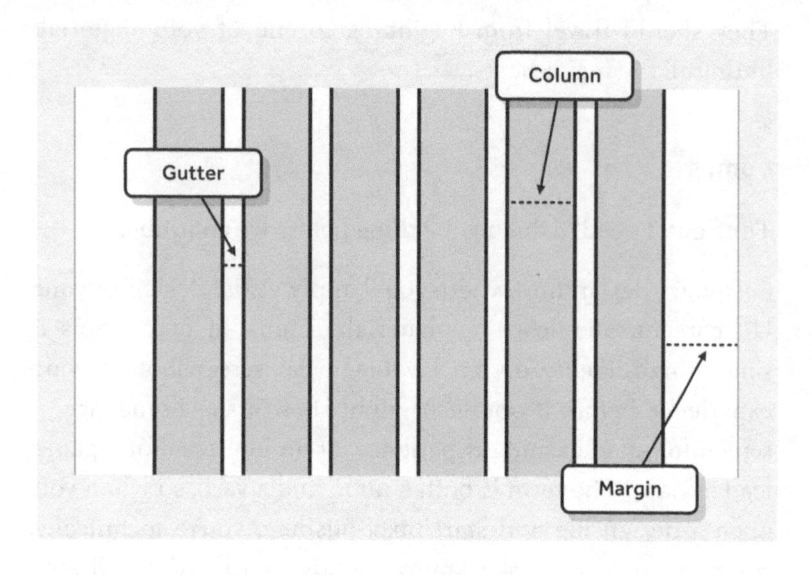

Remember ages ago when we talked about alignment? Me neither! But that's okay, because you can think of a *grid* as an alignment system for your whole project. It defines rules for positioning content and is used from design through technical development.

Modern grids have roots in editorial and print design, and they've evolved for the digital world. Like graph paper, you don't have to use every line, but it does make you think twice about slapping things down haphazardly. And while your grid might include both vertical and horizontal guides, the vertical (or column) grids will be your focus for most digital purposes.

We could take a deep dive here, but the point is this: enforcing an alignment system makes information easier to consume. Grids create guidelines to help the eye and the brain recognize structure. They ensure your bullets, sub-headers, questions, and elements are positioned consistently throughout your eLearning. They also save substantial effort when designing.

Here are a few grid-related terms worth knowing:

- **Margins:** The spaces around the outer edges of a grid.
- **Columns:** The main vertical containers for content.
- **Gutters:** The space between columns.
- **Rows:** Horizontal containers for content. These guides won't exist by default in most column grid templates, but as a designer, you'll define rules (double entendre!) in two dimensions.

Sounds great! Where can I get one?

You can always find grid templates or construct them yourself, but most design apps have built-in actions to generate grids for your comps. You'll supply the number of columns you want, the width of your canvas, and optional details of grid anatomy. Your software will then lay down a series of guides to match your specs. Poof! Instant grid system.

Technology is awesome, by the way. We used to do all that by hand.

When you move from design to development, your developer will also use a grid system, either coding one from scratch or leveraging a *front-end framework* that takes care of the details. It's your responsibility to ensure the grid you choose for design matches what they'll use for the build. Otherwise, you risk major changes after comp approval, and trust me, that's a bad day.

Responsive grids

While *responsive design* and *adaptive design* are topics for another book, keep in mind that as your digital experience scales up or down, your layout should adjust as well. And before we discuss how that affects your work, I need to walk through some semi-obvious concepts. Bear with me.

First, most eLearning will take place on a digital device and screen. Those screens come in various shapes and sizes: phone, tablet, laptop, or ultra-wide monitor. You might deliver your experience as a native app installed on an operating system in its own visual wrapper. Or you may deploy it online using code that's rendered in a browser.

If your experience lives online, the container for your experience (in this case, your browser window) is called a *viewport.* The viewport size can vary for lots of reasons, either at the hands of a user or based on screen size. When this happens, most modern sites or web apps are built to change, to avoid getting cropped or losing crucial functionality.

No surprises so far. We've all used the internet. But consider how this affects your design tasks. Your layout may need to scale or reshuffle as digital real estate changes. Something that works fine with three columns on a laptop may be illegible on your phone.

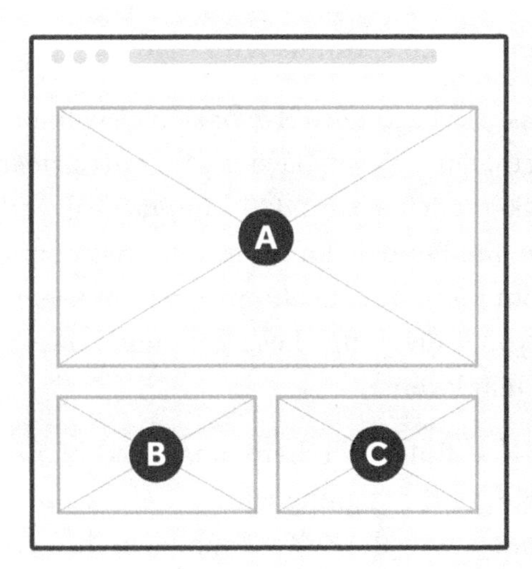

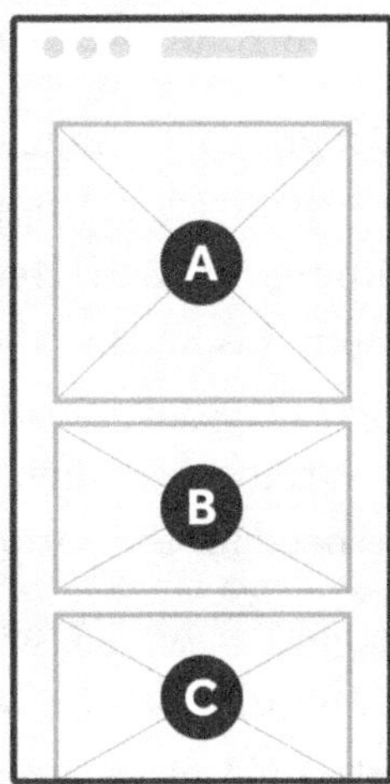

Also, the way content scales may differ, depending on how you want things to behave. Columns might constantly squash and stretch as you scale a browser window (a.k.a. fluid grid behavior) or snap to new dimensions at specified thresholds (a.k.a. *breakpoints*). Loosely speaking, this is the difference between *responsive* and *adaptive* behaviors, but don't worry too much about those. Just be aware of your options for scaling, decide how you want your layout to behave, and discuss these choices with your developers.

Finally, remember that you can't work at every possible size, so you're usually comping a few fixed widths. Essentially, you're creating snapshots of a moving structure. Responsive and adaptive design are about planning: how will things change, resize, or reshuffle as the viewport scales?

Keep in mind that not all eLearning authoring software allows for responsive design, so raise the topic when discussing your tech approach!

Do you want twelve or sixteen columns with that?

Most grids for online use have either twelve or sixteen columns for easy division. Does that mean every layout will show sixteen columns of text? Heck no. That would be impractical and, frankly, insane. But the number of columns in your master grid determines the maximum number of columns you can use. As you arrange content, you'll divvy up those columns however you need for the particular layout.

Imagine a screen with a full-width hero image positioned above three columns of text. If designing with a twelve-column grid, you'll place content across all twelve columns in that first row. Again: there are still twelve columns in your base grid, but they appear as a single column for your hero. In the next row, you'll use four grid columns for each text block, creating three visible columns of equal width.

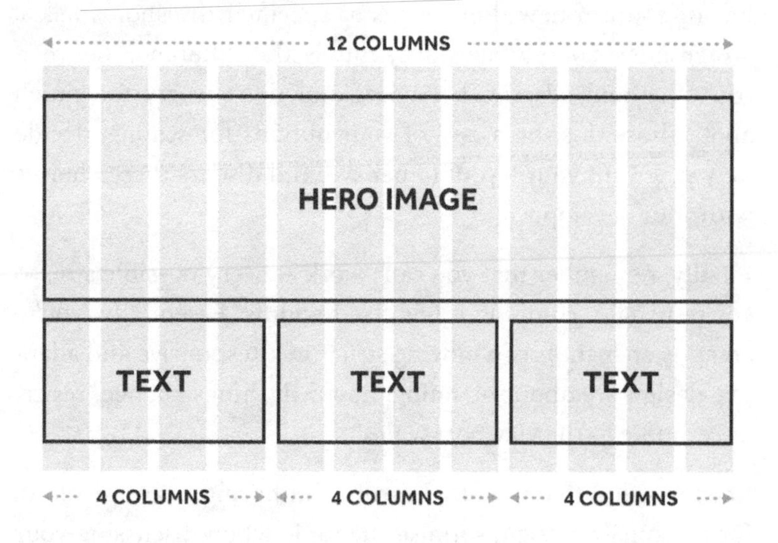

Although every layout uses the same grid, remember you'll usually be designing responsively. This means the width grows or shrinks depending on the size you're comping. While this varies by budget, staff, and company conventions, you'll generally produce at least one large and one small comp per screen as a reference for developers. These sizes are often called desktop and mobile comps, but that's just a convenient shorthand. With the growing spectrum of devices, dimensions, and screen sizes, what's rendered on one mobile device may vary from another. The structure can even change as you rotate from portrait to landscape mode!

A twelve-column grid is a common choice because most digital layouts work best using one, two, three, or (for extremely wide viewports) four columns. Twelve is divisible by all those numbers, making it easy to create columns of equal width.

If all this seems confusing, don't let it stress you out. Open your design software, generate grids for one wide desktop layout and one small mobile layout, and begin structuring content. You'll see where adjustments are needed.

Just remember two critical notes:

1. Choose a grid system that matches what you'll apply for development. Connect with whoever will write code, ask questions about scaling, and make sure you're aligned (hah!) on the number of columns.
2. Set your design grid before comping for each width. Make sure it's the same in every comp.

If you skip these steps and try to revise after producing dozens of comps, you're in for a rough ride.

Seven tips for typography

Good typography is an art and science all its own, but until your interest is piqued, let's save some time. Most primers kick off with a discussion of serifs and san-serifs, point out the nooks and crannies of font anatomy, and lead you on a meandering tour of the fifteenth-century printing press. But ultimately, you can start putting fonts to work with a few practical nuggets.

Enjoy this abridged version, and don't forget to tip your guide!

1. Know your key concepts, but don't get lost in the jargon.

- Typography terms can be nuanced and don't always have clean divisions. Sometimes, even the pros use them differently or incorrectly, and a five-minute web search will leave your head spinning with inconsistencies. Don't let that discourage or distract you. Learn the ideas behind the jargon, and focus on what you're trying to do.
- A *typeface* defines the core design for letterforms with common shapes and characteristics. *Helvetica* is a typeface.
- If a typeface represents the family name, a *font* is one specific child defined by weight and style. *Helvetica* is the font family. *Helvetica Bold Condensed Oblique* is a font. In some cases, size (e.g., 16 px vs. 24 px) is part of that definition (mostly due to the history of print I promised not to bring up), but this is less relevant today. Many designers use the terms font and typeface interchangeably, despite the distinction.

Miller Text Roman
Miller Text Italic
Miller Text Bold ← Font
Miller Text Bold Italic

Typeface
(a.k.a. font family)

- A *font family* (or type family) can have children of different weights or styles, also called *variants*. Practically speaking, font family is just another way to say typeface.
- *Style* is exactly what it sounds like, a font category based on shared visual characteristics, but it's a bit nuanced. Most people think of *serif* vs. *san-serif* as two primary font styles, but there are also *italic* (sometimes called *oblique*), script, blackletter, and so on.
- Font *weight* represents the thickness of a font. Some font names indicate their weight using words like *thin*, *regular* (also called *book*), *bold*, and *black*. These adjectives may be emphasized with words like *ultra* or *extra*.
- Some fonts are available in various *widths*, like *narrow*, *compressed*, *condensed*, and *extended*. Pro tip: never distort the width of a font in your design software unless you're working with a logo or another one-off custom treatment. Artificially scaling font width is considered bad practice.

2. Understand basic font spacing.

- *Tracking* (or *letter spacing*) is the default space between letters in a text block.
- *Kerning* controls the distance between two individual letters. Think of it as a fine-tuning override for tracking.

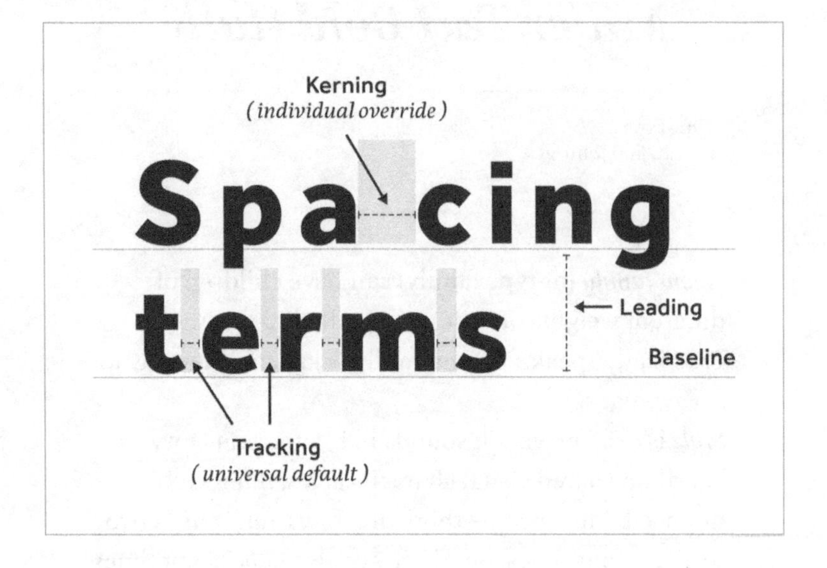

- *Leading* (pronounced "led-ing") is essentially line spacing—the vertical space between lines of text.
- As you transition into development, there are similar, but not identical, concepts for spacing. Occasionally they use the same names, but not always. It's worth noting that some settings from design software aren't easily reproduced in code. So, when working with a developer for adjustments, you may want to brush up on *cascading style sheets* (CSS).

3. Use *pixels* (px) as units when designing for digital.

- Your developer may eventually convert pixels to *rem* or *em* units in code, but don't worry about that—it's pixels for you! And never use *points* (pt) for eLearning. Those are for print.

4. *Desktop font* files are entirely different from *web fonts*.

- *Desktop fonts* are usable by your design software and appear in your comps. *Web fonts* are optimized for web use and displayed in a browser once your experience deploys online.
- Just because you have desktop fonts installed locally does not mean you have the necessary web fonts. To avoid last-minute changes or purchases, coordinate with your developer to ensure everyone has the assets they need.

5. Combine typefaces sparingly.

- Think of typefaces like spices; add them cautiously to avoid over-seasoning, especially when you're less experienced. Mixing fonts is one way to build contrast. It adds sophistication and flavor to your design but can quickly get muddled.
- As a rule of thumb, start with one to three unique fonts. If you wind up with more and you're not a seasoned pro, stop! Take a close, honest look at your work—there's a good chance you don't need all those extras. And when I say fonts in this context, I'm talking about style and family rather than pixel sizes. You'll almost certainly have more than three font sizes.

6. Use type to create contrast and hierarchy.

- In choosing your fonts, pay close attention to contrast and hierarchy. You can create both by pairing font sizes, weights, or styles that are different enough to notice. You might mix a large serif font for your title with smaller san-serif body text. Or combine a thick font for questions with thin fonts for answers.

funky

Sometimes you need to mix things up.

Mixing typefaces can add contrast to a design, making it easier to read. It can also help convey levels of hierarchy within your content.

- As you scan your layout, ask whether the reading order and parent vs. child relationships are clear.
- Adjust your contrast based on how much you wish to separate information. For example, if you need quotes to really stand out from your article text, you might create a drastic visual difference.

Van Helsing's early life was shrouded in darkness. Some believed that he was born in a small village in Transylvania, others in the Netherlands or Romania. What was certain was that he grew up in a world filled with danger and intrigue. As a young man, Van Helsing traveled extensively, seeking out the knowledge and training he needed to become a master vampire hunter. He studied ancient texts on the undead, learned the art of sword fighting and marksmanship, and even delved into the world of alchemy and magic.

Van Helsing's reputation as a vampire hunter soon grew, and he was sought out by those who needed his expertise. He was hired by wealthy aristocrats, desperate townspeople, and even governments to investigate and eradicate the undead. It was said that he had a particular talent for identifying the telltale signs of vampirism that others could not. He was able to identify vampires by their physical characteristics, behaviors, and even their smell. His methods may have seemed unorthodox, but they were undeniably effective.

Van Helsing's career as a vampire hunter reached its peak with his confrontation with the infamous Count Dracula. The two clashed in a battle that lasted for days. Van Helsing used all of his knowledge and skill to hold his own against the count, and ultimately emerged victorious. Dracula was destroyed, and Van Helsing became a legend. His name became synonymous with vampire hunting, and he was

“

Garlic, stakes, holy water–you name it, I've tried it. When dealing with vampires, sunlight works best!

\- ***Van Helsing, Influencer***

It was said that he had a particular talent for identifying the telltale signs of vampirism that others could not. He was able to identify vampires by their physical characteristics, behaviors, and even their smell. His methods may have seemed unorthodox, but they were undeniably effective.

Despite his
He rarely
that he had
knowledge
supernatur
extraordina
rumors as
that there

In his later
and other
him change
science, usi
techniques.
sharing his
that he died
man who h

Van Helsing
of vampire
have been
effective. He
against the
and an unw
lived, a myst
world of sup

Keep an eye out for my next book, Instructional Design for Vampire Hunters!

7. Be mindful of accessibility and legibility.

- You may be tempted to use small font sizes for your paragraph or UI text, but remember you'll sacrifice readability, especially for those with impaired vision. Accessibility compliance is crucial and has all kinds of implications. Generally, I'd recommend a minimum font size of 16 px for body copy when working online. But do your research and define your requirements.

What do you mean I'm skipping a step?

But what about wireframes, you say? Good point. There's usually an intermediate step between content planning and visual design. The process of sketching layouts before diving into comps is called *wireframing*, and it's a core activity in information architecture and user experience design.

Wireframes are un-styled diagrams of an interface using raw lines and boxes. They let you focus on placement, structure, and functionality without getting distracted by visual design.

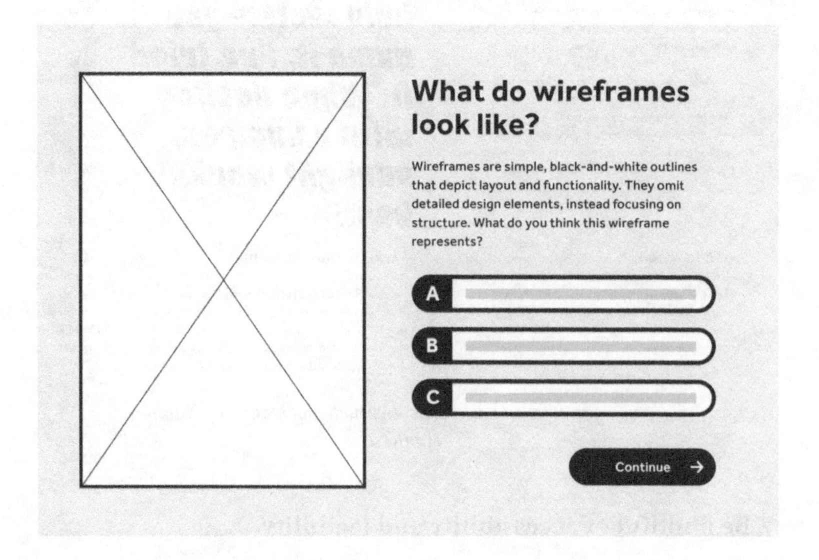

I don't want to derail you with yet another technique, but let's say it's worth sketching layouts in advance. Before locking down colors, fonts, and visual personality for UI, decide if you want that UI in the first place!

Chapter takeaways

I'm a designer by trade, so I get chatty about this stuff. Personal passion aside, I wrote this chapter because visual design principles are potent and practical. Even ten minutes of foundation can improve your design game and create a cleaner, more professional experience for your learners.

- Good visual design builds upon long-established principles. These principles help you organize information, reduce wasted effort, and direct attention where it's most important.

- *Comps* are design files where you'll apply visual design to your UI and content layouts. You'll use them to share work with teams, clients, and stakeholders. Generally, you'll produce batches of comps to gather feedback and make revisions before development.
- *Grids* are vital tools to enforce alignment across your comps. You'll generate grid systems in your design software and use them consistently throughout your project.
- Understanding grid anatomy is useful for design planning and communicating with your developers.
- Twelve or sixteen-column grids are standard for easy division, but whatever you choose should align with your final build.
- When designing responsively, it's essential to consider how your experience will scale and adapt as your viewport changes.
- Typography is a deep branch of visual design, but you can begin with some practical basics. You'll use type to establish contrast and hierarchy. Combine and add font families sparingly.

15

CREATE A STYLE GUIDE AND LAYOUT TEMPLATES

Consistency is the bedrock of learning. If a word's meaning, spelling, or pronunciation changed each time you used it, no one would understand you. At the very least, they'd waste loads of effort decoding your request for a topato chip. I mean, an opotat ciph. I mean... JUST GIVE ME ALL YOUR PICH PATTOOS!!!

Tiring, isn't it?

Similarly, if your help icon or continue button changed color, language, or screen placement, users would burn energy reorienting themselves rather than focusing on content.

Learners have enough to do when absorbing new information. Consistent design preserves their energy by keeping your interface familiar. It helps them focus on the artwork, not the frame.

In this chapter, we'll use some of the design principles we've covered. You'll create a style guide to define visual rules, then build layouts for reuse. This streamlines your work, locks client approvals, and keeps your learner experience consistent.

But first: what the heck's in a style guide?

Create a style guide

You'll see all kinds of *style guides* across companies and industries—writing, branding, and visual design being the most common. Some overlap in purpose, and others are unrelated.

Here, we're talking about a reference for design, typography, and interface elements used to create a consistent aesthetic. A style guide lays out visual rules for anyone designing your learning experience and is especially helpful when people are working in teams.

Usually, I'm a bit of a stickler about terminology. You'll also hear terms like UI kit and design system in the mix, and definitions vary. Here's a quick overview to avoid confusion.

- A basic *visual style guide* provides fundamental guidance and establishes a baseline for design rules.
- A *UI kit* is a specialized, tactical artifact focusing on user interface elements. It defines content blocks and components and serves as an intermediary between designers and developers.
- A *design system* is a more robust visual UI library that defines styles for all known components at multiple levels. It's used for large-scale projects or as a repository of company-wide materials. You might think of style guide vs. design system as the difference between grade school and grad school.
- An *editorial style guide* keeps copywriting and messaging consistent across platforms. It's essential for determining voice, personality, and word usage.

To be honest, UI kit is a more accurate term for what you'll be doing here, but for simplicity, we'll call this your straight-up style guide for now. When you reach the point of quibbling about design semantics, you've leveled up!

What should your style guide cover?

Your style guide can be as in-depth as you'd like. While you want to leave *some* room for creativity, the more it covers, the less you need to interpret or explain to others. But at minimum, it should include the following items.

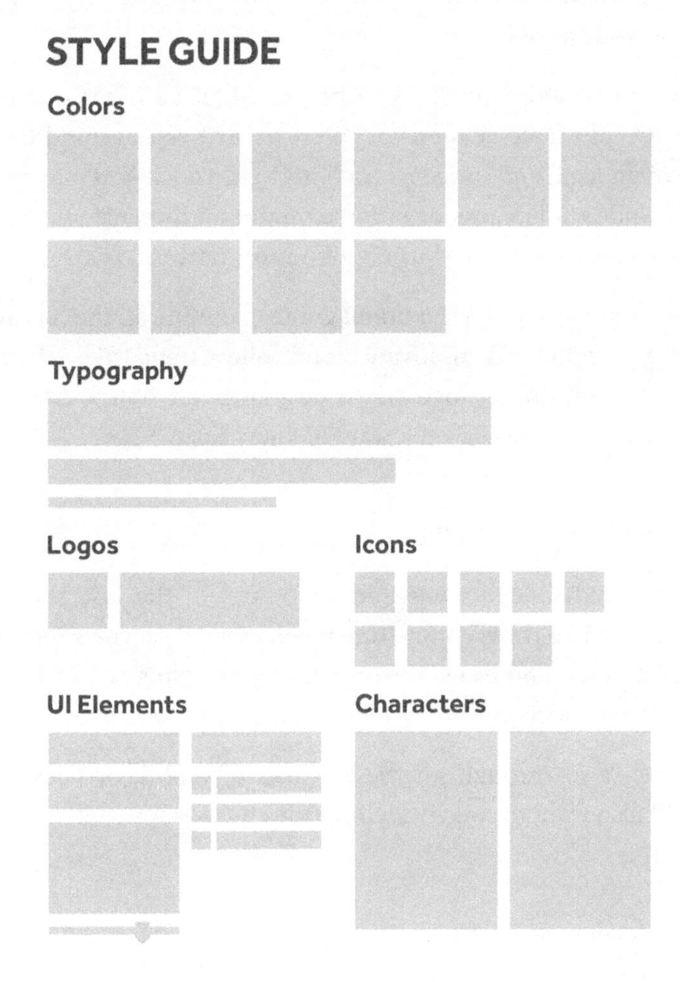

Logo treatments

Logos (also called lockups) are some of the first items to add to your style guide because 1) usually, there's not much you can do to change them, and 2) all your other elements should play nice with your logo from a design perspective. Logos are part of a company's identity, and most of the time, you can't tweak them to suit your needs.

How and where will the logo be placed? What sizes can you use? Do alternate versions or colors exist? Keep in mind there are often legal and usage restrictions to consider. You may have to include trademarks or enforce minimum dimensions based on viewport size or platform (mobile app, browser, etc.).

Clients sometimes have a brand guide to outline all the dos and don'ts. Ask for this in advance and follow their rules. If they require a minimum logo width of 400 px for online use and forbid color changes, you'll want to know early.

Font styles and contexts

We covered font families and properties in the last chapter. Choose one to three contrasting typefaces and set rules for size, weight, color, and usage. Inconsistent font styling is a common problem, especially when there's a larger team involved.

We won't go through all the nuances of HTML or CSS, but you'll also want to specify styling rules for various contexts. For example:

- Headers of all levels
- Body copy
- Rich text
- UI text (button and label text, etc.)

Be sure to plan for responsive design if you're deploying online. This allows text to grow or shrink as your browser scales.

Color palette with hex values

Most brands create color palettes for digital use. Whether these exist already, or it's part of your job to define them, make sure your style guide includes swatches and hex values for each color. RGB is also fine, albeit less common. CMYK and Pantone values are for print only. If new colors come up during design, add them as you go, and be sure to prune any you're no longer using.

Common interface elements

Since this is for an eLearning experience, many interfaces can (and should) be repurposed. This includes:

- Buttons & links
- Form elements (fields, labels, drop-down menus, validation text, etc.)
- Navigation elements
- Icons (social media, navigation and help icons, etc.)
- On-screen alerts
- Modal windows (a.k.a. popups or overlays that appear above the main screen)

Also, consider interactive UI for quizzes, practice activities, or games you'll design.

Grids and spacing

We covered grid basics in the last chapter, but it's worth a quick recap.

Your grid provides literal guidelines for alignment and spacing. Twelve or sixteen columns are common in responsive design, so pick a system for your project and stick with it. Your choice will establish margins, gutters, and default spacing between elements.

Most visual design software has presets to generate your grid system. However, a word of advice: when building a custom eLearning experience, talk to your developer before starting.

During technical development (a.k.a. coding), your developers may use a pre-existing grid framework, and their requirements will override any choices you make in the design stage. If this is the case, I recommend matching their grid in your style guide. Either way, you'll want to chat with them beforehand.

Create common layout templates

Designing comps is one of the most time-consuming things you'll do as an instructional designer. Fortunately, you can save lots of time, effort, and client budget by putting your new style guide to work and creating common templates.

As mentioned earlier, a comp is a polished visual preview of a page, screen, or composition. For easy reference, we'll call the single unit comp for an eLearning layout a slide. In other words, the first screen a user might see could be the Introduction or Title slide. This comp might include a primary title, description, background image, and start button.

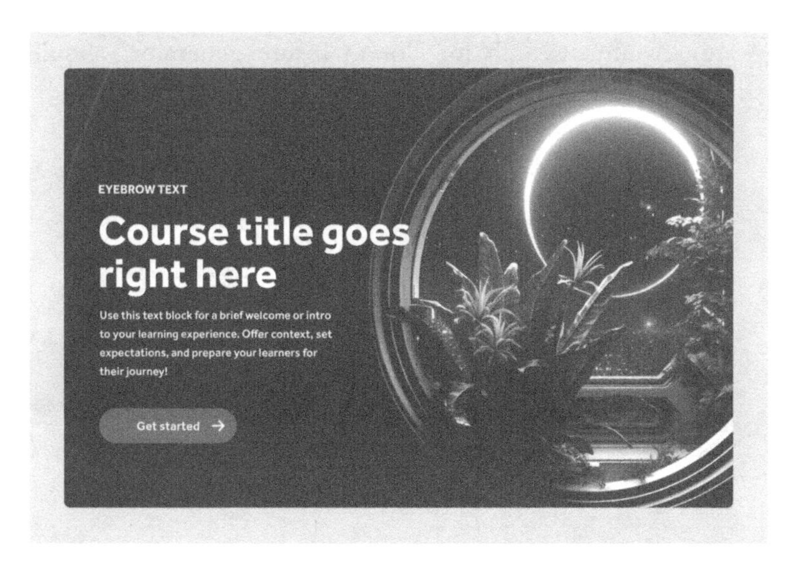

The view from our space garden!

And yes, sometimes there's a bit of a chicken and egg situation. You might first design a complete slide to decide how the UI and spacing should look, then add individual elements to the style guide. But if it wasn't originally in the style guide, how did you add it to your comp?

You don't need to overthink this. Just ensure your design is consistent throughout the experience, and keep your style guide up-to-date. Beyond that, do what you gotta do.

What about the text?

For long-form text, add placeholders to approximate the amount of content allowed per template. Designers often use *Greek text* for this purpose—a long-held print convention to insert gobbledegook text which, despite the name, is not actually from the Greek language. This helps clients assess content length without being distracted by messaging.

A quick online search for "lorem ipsum generator" should supply plenty of references.

We used to see who could memorize the most lorem ipsum, because that's what happens when you're comping at 3:00 a.m. every Sunday!

Which layouts do I need?

Based on your storyboard, write a list of layouts you know you'll use, especially those that appear repeatedly. You'll design one template for each slide in your list, then swap content as needed.

Rather than creating unique comps for every screen in your project, you're reducing work and establishing conventions. When a user sees a particular layout, they'll begin to associate an activity or message. Layout A always marks a new chapter, while Layout B prepares them for a quiz. Some standard layouts include:

- Title slide

- Chapter or section divider
- Main message with bullets
- Multiple choice
- Result and feedback
- Question
- Character discussion
- Section summary

It's worth noting that some eLearning authoring tools ship with pre-defined template libraries. You'll still want to customize them for your needs, aesthetics, and clients, but using these as a starting point can save considerable time.

Share your stuff

These steps will save you plenty of effort, but they're not just for you. Before going too far, now's a good time to share your work with clients and stakeholders.

Presenting your style guide and layouts invites early feedback and helps confirm your design choices. Use this step to make changes before producing every screen. It's a fast and nimble approach to get clients on board with a sample.

One more tip: avoid presenting your style guide in a vacuum. Some clients aren't design-savvy, and the response can often be: "I can't approve anything until I see the finished product. Why am I looking at buttons and colors out of context?" Showing the style guide applied to at least a few comps is much more effective.

Tools of the trade

Designers use many tools for style guides and templates, but as I mentioned earlier, these preferences shift with the industry winds. Your software will vary by company or situation. Each offers different benefits, specialties, pricing, and features.

However, digital tools are subordinate to your main task: to define clear visual rules and enforce consistency throughout your project. Doing so streamlines your work and creates eLearning that keeps learners focused.

Chapter takeaways

Recording design rules up front can save you and your team all kinds of headaches. While it requires investment, this step lets you collect early client feedback, reduce effort for your learners, and build consistent, reusable assets.

Below are some key concepts worth remembering.

- Visual consistency sits at the heart of communication. It sets expectations, preserves mental energy, and helps learners focus on your material rather than your interface.
- A *style guide*, as defined for our purposes, is an artifact that outlines visual rules for design. It defines colors, typography, and interfaces, and enforces consistency between team members.
- Creating reusable templates can save time during the design process. Here, you'll apply rules from your style guide to common layouts.
- Share results with stakeholders once you've developed a style guide and a few representative comps. This lets you validate work and make changes before going too far.

16

PROTOTYPING

We'll define a *prototype* as any mockup of content, features, or interfaces made to evaluate something without building the final product. In this phase, you'll flesh out just enough to get feedback without burning all your budget. Sounds promising, right?

Prototypes can be hugely rewarding for both designers and stakeholders, but it's vital to match your execution with the goal. Are you soliciting feedback, testing versions of content, or evaluating activities? Decide why you need a mockup, then customize your approach.

The first reason to prototype

In eLearning, prototypes often follow storyboards as the next client deliverable. While you might use placeholder content, prototypes bring the narrative to life and let stakeholders explore your experience as a whole.

For this purpose, you'll populate layouts with content from your storyboards and build a clickable preview. The objective is to get client approval on your visual design, story beats, and overall user flow before moving to development.

Targeted prototypes

When clients see static comps, they may not fully realize what you had in mind. And if they don't understand how something functions, your work may pause until they do.

In addition to helping you move beyond storyboards, here are a few places where prototypes come in handy.

- To help clients visualize scenarios or materials that are difficult to describe verbally.
- To explain interfaces that have multiple states, such as responsive navigation, sorting or filtering systems, and multi-step games.
- To test different versions of activities or content with users.

You may also encounter common hiccups or roadblocks. For example:

- After reviewing flat comps for an exercise, clients still don't understand how your UI will function.
- You want feedback on an animation style before applying it everywhere.
- Clients won't approve a complex scenario through storyboards alone.
- Clients are getting antsy in the gap between storyboards and final development, and you want to show progress without derailing your team.

Some clients have more imagination than others, but sharing complex work is always a challenge.

Avoid precious ideas and the Grand Reveal

If you loved a particular idea while brainstorming and it falls flat when you build it, don't be afraid to scrap or revise. That's the whole point of prototyping. Likewise, if you spend too much time polishing for a Grand Reveal at the end, you risk wasting energy and budget on something ineffective.

Sure, you'll find formal diagrams to show you exactly when to do what, but basically:

- Invest some work in an idea.
- Build a prototype for presentation or hands-on testing.
- Share it with relevant people or groups, then observe or gather feedback.
- Use insights and revise.
- Repeat as necessary.

In addition to helping you course-correct along the way, prototypes give your clients a sneak peek without huge surprises at the end.

Oh! And be sure to add this time to your project plan. Done and done.

Prototyping workflow

It's easy to go overboard once you've finished comps and storyboards. You may want to build all the bells and whistles in prototype form. But if you spend 99% of your budget building a high-fidelity prototype—and you're not working with your final software—you won't have much left for development.

How do you figure out what to include in your prototype, how much time to spend, and how detailed it should be?

1. What answers do you need from your audience?

One goal of a client-facing prototype is almost always to get your stakeholders excited by seeing the material brought to life. After all the long interviews, working sessions, outlines, and storyboard reviews, they'll finally get a sense of what the experience will feel like.

But beyond that, define specific features to validate or test. Do you have a list of questions you need answered? Are you unsure if a tricky chapter will confuse your audience? Are you testing different versions of an interface? Or are approvals blocked because your client can't envision the final product with images alone?

Decide where to focus, then invest detail to get your answers. Leave the rest as low-fidelity as possible. For example:

- If you're testing a new quiz interface to ensure it's intuitive, you may want to build a high-fidelity UI mockup to see how it works with users. There's no need to get bogged down by copywriting or imagery because that's not what you're testing. Just design enough to set the stage.
- If you're evaluating a scenario to see if it feels long, you might create a click-through of your longest sequence. In this case, you might not spend time on the interface. Here, you're focusing on the narrative, not the navigation.

Some questions to guide your scope:

- Which content, sections, or scenarios should you mock up?
- Do you need to demonstrate multiple paths, or can you make an on-rails linear experience?
- How much interactivity, animation, and interface detail do you need to accomplish the goal of your prototype?
- Can you build this yourself, or do you need help from others?
- Will your audience get distracted if something is low fidelity?
- What impact would higher fidelity have on your timeline and budget?

Remember that high-fidelity prototypes take longer, so reduce scope when increasing detail. In other words, if you're building something complex, be sure to make less of it.

2. Pick your prototyping tool

Remember when we brought this up in the *Get organized* chapter? Well, here we are!

If you're building with an eLearning authoring tool (see *Tools & Resources* for examples), you might prototype directly in the software you'll use for the live experience. This can save time and lets you reuse work later.

However, when relying on custom development, you may need other tools here. For example, if the goal of your prototype is to test whether a scenario feels long, you might use software that just links between slides. You've got plenty of options for this, but the process is quick and dirty: export flat images for each state, add some links, and boom! You've got a clickable prototype to test paths through a storyline.

On the other hand, if you need to model video, media, and interactivity, you'll probably want something more robust. Such tools usually have a higher learning curve, but that's not always a negative.

As mentioned in earlier chapters, I believe there's no one-size-fits-all prototyping software. Even if your company only uses Software X for its mockups, new tools pop up constantly. Need to learn something new for the best presentation? Go for it!

That said, your personal experience, workplace standards, and comfort level all come into play. A basic tool you know fluently will help you move faster than unfamiliar software. Balance the pros and cons, then make your pick.

3. Set audience expectations before presenting

Repeat after me: "This is a prototype to demonstrate X with the purpose of Y. This is not the finished product."

Whether your prototype is for clients or learners, setting expectations is crucial. Without context, your audience may assume everything they're about to see is clickable, coded, and polished. Constant interruptions asking why this or that isn't working can kill your momentum and their enthusiasm.

Here are a few tips for setting the stage.

Make the limits of your prototype clear. Explicitly call out what's included and functioning. Try to anticipate questions that might derail your presentation. For example:

- "You'll see sketches or stock images instead of the final video."
- "You'll see on-screen text but won't hear voice-over."

- "Only one path is clickable to follow a specific user flow."
- "You can explore the entire branched scenario, but we haven't added animation."

Describe the feedback you're hoping for at each step. Give them examples of helpful and irrelevant comments.

- "Focus on interactivity and UI rather than specific copywriting. Feedback like 'We'd like more choices for each quiz' or 'We need a way to show users how much they have left' is perfect."
- "We'll discuss animation style in a later review. Today we want to focus on the narrative."

Use low-fidelity assets to make placeholders obvious. If you want a client focused on text and don't want them distracted by images, you might use sketches instead.

Surprise your audience with something unexpected. Animate a static feature, include content from your next sprint, or add easter eggs to the experience. After all the documents and discussions, prototypes let you bring the experience to life. Don't be afraid to exercise showmanship!

Your goal here is to excite the audience, set expectations, and prime them for relevant feedback. Helping them understand what you've built—and what you haven't—goes a long way.

4. Present, get feedback, and (if appropriate) test with users

Once your prototype is in good shape, share it with your clients, stakeholders, SMEs, or testers, and start collecting feedback. Early reviews don't have to be complicated. Set aside time with your audience. Then decide how much you'll walk through and where they can explore on their own. Just remember, sending unaccompanied links can be risky.

By definition, a prototype shows work in progress, and clients can get distracted when they're unsure where to focus. You might email a link asking for animation notes, only to find them confused by inactive search fields or placeholder text. Your cries of "But we haven't finished that yet!" won't reclaim wasted review time.

In most cases, I suggest presenting instead of sending links and hoping for the best. It allows you to answer questions and point out the unfinished bits. When real-time presentations aren't an option, be sure to set the stage carefully. Clients won't always read your disclaimers.

5. Apply what you've learned and iterate

After gathering input from your audience, you'll need to decide how to use it. That means either iterating on your work or progressing to development.

Every revision requires another round of presentation and testing, so factor that into your project plans. While some companies use agile timelines that cycle indefinitely, you can't spin forever on a fixed budget. At some point, most projects need client approval to move on.

Chapter takeaways

Prototypes can be an exciting way for clients to see their discussions, goals, and materials brought to life. They offer a preview before diving into development, and help pinpoint what's working and what needs a tweak or two.

- *Prototypes* help clients visualize complex materials, activities, or interfaces that are difficult to explain with comps alone.
- Assign each prototype a purpose and level of fidelity. Everything you mock up should justify your investment.
- Let the goals of your prototype dictate your software choice rather than the other way around.
- Set client expectations before sharing. Your audience should know what's working, what's not, and the kinds of feedback you need.
- Gather feedback and use it to iterate. Prototypes help improve your eLearning and get client approval before building the final product.

17

ADD INTERACTIVITY AND MULTIMEDIA

Imagine writing the most valuable training ever created, then just throwing it on-screen with skyscrapers of text and a few playback controls. Voilà! It's digital, so now it's eLearning, right?

Yeah... not so much.

You may have poured your soul into content, but despite all that effort, there's a likelihood your course will still 100% suck. Quality copywriting is critical, but unless you're hosting a blog, raw text can't do all the work.

Interactivity and media let you mix things up and use bits from the instructional design toolkit. Rouse the brain with movement and variety. Engage learners in active roles rather than anesthetizing them with passive, monotonous lessons. As we learned in *Designing for memory*, interaction helps people absorb your material.

From a design perspective, this can be a super fun part of the build process. It's your chance to enrich content with compelling motion and storytelling. But all that power should be carefully channeled, and there's a fine line between action and distraction.

Mayer's 12 Principles of Multimedia Learning

Psychologist Dr. Richard Mayer developed twelve insights to describe how people learn from multimedia (Clark & Mayer, 2016). Look closely, and you'll see familiar ideas from visual design and designing for memory.

1. **Coherence:** People learn best when you cut out the noise. Whether on-screen text, audio, or design, find ways to reduce and simplify your content. If it doesn't benefit your learners, take it out.
2. **Signaling:** Show your learner where to direct their attention. Cluttered screens distract people from the important bits. Show them where to focus by bolding and highlighting words, using animated arrows, and breaking large chunks of text onto different slides.
3. **Redundancy:** Duplication can be distracting, so streamline your content. For example, you might not need supplemental text if you already have visuals and voice-over. Granted, this can be tricky when designing for accessibility.
4. **Spatial Contiguity:** Spatial Contiguity is just a fancy way of saying proximity, which, I guess, is also kind of fancy! Regardless, people learn best when related assets sit close together. When arranging content, group text and graphics that relay the same idea. This reinforces a message and avoids attention splitting.
5. **Temporal Contiguity:** Make sure your audio and visual information are in sync. Simply put, use voice-overs to complement what's happening in your

animation at the same time. Don't talk about one topic while showing something unrelated.

6. **Segmenting:** Long streams of information become marathon sessions, and learners don't absorb as much as they should. Break your content into digestible segments or chunks.
7. **Pre-Training:** Fundamentals are essential. Make sure your audience grasps the basics of a topic before moving deeper. Present core concepts and vocabulary up front, then summarize to connect the dots.
8. **Modality:** When telling a story through graphics, it's sometimes better to reinforce them with descriptive audio rather than on-screen text. Merging two streams of visual information (e.g., text and images) may be overwhelming, but the brain can process audio and visuals separately. You wouldn't watch a movie while reading a book, right? But a movie with sound works just fine. However, this principle comes with some nuance and practical hurdles. So rather than ditching on-screen text altogether, consider this a reminder to balance your content and avoid over-taxing any one sensory channel.
9. **Multimedia:** Support your words with corresponding visuals. Instead of showing a wall of raw text, use visuals to break up your content and illustrate key points.
10. **Personalization:** Humans don't connect emotionally with stodgy, formal, or academic language. Writing in a conversational voice helps learners relate to the narrator. How you do this varies by audience—what's comfortable for one group may be off-putting for another. (That's why you should know your learner.)

11. **Voice:** People prefer real human voices. Record your audio from bonafide humans, or make sure your text-to-speech is indistinguishable.
12. **Image:** This one's a little controversial, given the success of modern tutorials and educational videos, but here we go. The *Image Principle* proposes that videos of people talking (a.k.a. talking head videos) may not always be the most effective teaching method. Consider using visuals that illustrate and demonstrate a concept rather than focusing solely on the instructor.

There you go—twelve handy principles to guide your eLearning media. Use them for inspiration as you design and script your content.

Next, let's talk about the hows and whys of letting users click, drag, and move what you give them.

Using interactivity to improve retention

Is there a difference between lifting weights yourself and watching someone else do it? I wish there weren't, but reality is cruel. You've gotta put in the work to build those muscles.

eLearning relies on the same idea. For learners to build knowledge and skill, they need to interact with the material. Fortunately, digital platforms are made for just that. But how much participation will you expect?

Four levels of interactivity and their implications

As you may have noticed, instructional designers love classifying things, so let's roll with it! They often sort eLearning into *four levels of interactivity* based on content structure and depth of participation. Definitions vary, but here's the general breakdown.

1. **Passive interactivity:** In an analog setting, *passive interactivity* equates to sitting in a classroom, listening as your instructor recites information. No raising your hand, no exercises. Hence, passive. Digitally speaking, users can usually navigate or click through slides using basic controls—fine for simple materials that don't merit a deep dive. But for practice or decision-based tasks, you may want to look elsewhere.
2. **Limited interactivity:** *Limited interactivity* builds on the previous level, adding quizzes, drag-and-drop activities, rollovers, and interactive animation. Moving to this level lets users test and apply the material to some degree. It opens the door to matching, sequencing, and asking questions, and makes an experience more engaging for learners since they actually get to do something.
3. **Moderate interactivity:** *Moderate interactivity* takes participation a step further. This level often introduces storytelling, and users may need to solve problems by manipulating on-screen content. Interfaces and interaction patterns become less rigid. Animated videos, gamified content, and branched scenarios are all in-bounds. You'll encourage learners to immerse themselves, think creatively, and work on deeper problem-solving.
4. **Advanced interactivity:** Definitions for *advanced interactivity* change as the notion of advanced evolves. But at its core, this tier involves more realistic simulation. Interactive video scenarios, AR or VR exercises, high-fidelity environments, and other immersive content falls into this bucket. By mimicking

reality as closely as possible, learners practice skills under safe conditions. This is a great option to familiarize them with a setting or have them react under pressure. Obviously, such bells and whistles require more work. This level is also called real-time, simulation, or full-immersion interactivity.

As you dial up interactivity, you ask learners to exert more effort. Intricate games, puzzles, and interfaces can help learning stick, but they're also mentally taxing.

Look at your experience as a whole. Have you assembled a healthy balance of shallow and deep activities? Did you consider cognitive load for each session? Are you giving learners a chance to breathe?

An action movie with end-to-end explosions has no highs or lows because everything's the same. It's dull and exhausting, and audiences quickly grow numb from the lack of variety. View your eLearning as storytelling, and mix up the energy.

Animate with purpose

If you're investing in animation, you should first consider why. A smidge of planning will help you decide where to add spice and direct attention. Here are a few good reasons to animate.

- "This slide content is dull and needs something to wake the audience." Good reason.
- "We should draw attention to this collapsed help icon so users know it's there." Also, a good reason.
- "We're showing a process humans can't normally see." Top-notch rationale, there.
- "We want to clarify where you're going and where you're leaving." Yes, please!

Psychologist Richard Lowe (2004) cites two main reasons for using animation.

First, there's *affective purpose*—in other words, capturing attention. You might animate to introduce humor or make dry material more spicy.

The second reason is *cognitive purpose*. Motion can tell a story that would be difficult to explain otherwise. You may need to show a complex process, present something invisible to the eye, or mark steps that occur over an impractical time span. In this context, animation comes in handy for illustrating change.

Navigation and progression

I'll expand on that second reason, especially as it relates to navigation. You might use animation to establish *direction*, *origin*, and *timeline*. In other words, where did you come from, where are you going, and how far have you gone?

Origin might apply if you're opening and closing a menu or overlay by clicking an icon. (Click the icon, and the menu scales up from within, moving from nothing into existence. Click it again, and the menu gets sucked back home.) If you imagine such a transition without animation, not only would it feel jarring, but the mind has to take an extra step to interpret where the thing came from and where it returned. Animated transitions smooth the mental path between opened and closed and make your experience more seamless.

Similarly, an example of *direction* and *timeline* is when you click a right-facing arrow and content slides off-screen to the left, instantly communicating your progression *forward* in time. This motion is so natural it barely registers. It's a given for presentations, books (although the direction swaps in some languages), and page-to-page content.

Movement is also useful for storytelling. It clarifies where you're moving to and where you've come from. As interactions and branches grow more complex, these tactics become critical for orienting your users.

Investing in motion

Animation can impart knowledge in ways that infographics, text, and static images cannot, but it also burns time and budget. The style, subject matter, and execution all impact the talent required to pull things off, so choose your methods carefully. For example, the resources needed for hand-drawn vs. 3D animation are entirely different.

Even with an unlimited budget, you should always ask if motion enriches your content or detracts from its message. Animation should enliven or explain. Not everything needs to dance all the time.

Surprise and delight

There's a fine line between establishing patterns and lulling people to sleep. Don't be afraid to break up your training rhythm. Adding surprising, funny, or delightful touches can give the brain a break from dense material and entertain your learners in places they might not expect.

These surprise-and-delight moments needn't be drastic. You may want to blur backgrounds when users launch help text, reward success with animations, or add life to your menus. Get creative and have some (material-appropriate) fun with it.

Adding thoughtful details rouses your learners and makes eLearning feel richer. Your audience will appreciate the effort.

Chapter takeaways

Interactivity, media, and animation can spice up your eLearning, but use them wisely! Add bells and whistles that enhance your materials rather than distract from them.

- *Mayer's 12 Principles of Multimedia Learning* set guidelines for multimedia instruction. They offer best practices for audio, visuals, animation, text, and content.
- Instructional designers use *four levels of interactivity* to classify an experience's depth. Activities should align with your material, and each tier comes with costs and benefits for both learner and designer.
- Motion can enhance eLearning by improving UI, illustrating complex processes, and adding life to your content. Just make sure you have good reasons to animate, and plan for the investment.

18

DEVELOP AND TEST

By this point, you've got a high-fidelity storyboard. You've designed, prototyped, and revised key moments from your eLearning experience. All those steps helped you lock client approval where necessary. Now it's time to start building!

Many of your build specifics were laid out in the *Get organized* chapter, particularly who's handling technical development. Maybe that's you, an eLearning developer, or a team of web developers.

Regardless, here you'll focus on the final product. You can still share progress and make changes, but everything should reflect the live experience—code, assets, visuals, and copywriting. Any placeholders or previews will be short-lived. And before going live, you'll ensure everything works as intended, without bugs or errors.

Whether you're building with authoring tools or custom code, we'll look at the pain points most projects have in common. As the instructional designer, it may be your job to coordinate everyone. So let's dig in!

The development process

Early on, we looked at different ways to build an eLearning experience. One method is to work within authoring software, creating content with whatever components, templates, and features it offers.

In this case, you likely used the same tool for prototyping. But now you're finalizing details and content. When going live, you push files to an LMS or dedicated server.

This step demands more collaboration if you're working with devs to write custom code. You'll hand off storyboards, answer questions, elaborate on style guides, and give feedback as the experience takes shape.

And on that second road, you'll need some basic lingo to avoid sounding noob-ish with your devs.

- A *repository* is a central location where code and other files are stored and managed.
- *Deployment* is the process of releasing digital products to their intended audience. Generally, you can think of this as pushing code from one place to another.
- A *production environment* is where live code is deployed for users. This contrasts with *development* and *staging* environments used to build and test code before pushing it live.
- *Remediation* involves squashing bugs and fixing issues.
- *Versioning* is the process of tracking product releases over time.

- The *front end* describes parts of a digital experience users interact with directly, like the content and UI.
- The *back end* refers to systems that process data behind the scenes. It handles databases, user authentication, and requests from the front end.
- *Content management systems* (CMS) are used to upload, create, edit, or publish digital content. They are often managed by non-developers.
- *APIs* (Application Programming Interfaces) define how different apps can interact. For example, they may set rules for how your eLearning communicates with social media or an LMS.

Uploading content and media

As your project solidifies, you should learn the media workflow. Are all your assets exported, optimized, and ready to post? Who will upload media and written content? And where does that happen?

During the build phase, you'll output final text and assets, then place them wherever they need to go. You may upload them to a CMS (described above) or add them through your authoring software.

Regardless, you'll work at a higher level of polish now. You've moved beyond experimentation and are pushing toward launch.

Speaking of which...

Make sure everything works as intended

This book isn't a treatise on the technical development process (thankfully). But it's worth some due diligence to ensure your eLearning is tested and bug-free before sharing it with learners. Two standard practices can help.

- *Quality assurance (QA) testing* is the process of validating that an experience meets your original requirements. Ideally, it's conducted by an independent team (people not involved in the build) and often follows a test plan.
- *User acceptance testing (UAT)* is a lot like QA, but considers the user perspective (your learners, in this case). Products are assessed from a real-world, human point of view rather than checking off specific requirements.

Few digital experiences are flawless out of the gate, so be sure to budget for testing and fixes.

It's also worth noting there are many levels here. Sometimes companies have an entire team dedicated to QA. Sometimes it's left to developers and project managers for informal testing. Your approach depends on budget, timeline, resources, and the risks if something blows up, figuratively speaking.

This process can range from:

- Testing yourself.
- Recruiting coworkers to find issues, especially those unfamiliar with the work.
- Sending your experience to a QA team for full review and bug reporting. This is usually followed by remediation (a.k.a. fixing technical issues) and another round of QA.
- Engaging third parties to assess specific requirements, from optimization and performance to accessibility compliance.

At the very least, it helps to have a few people go through your eLearning with fresh eyes. So in the spirit of job aids and required information, here's a list of tips for your Bug Hunters!

- Try to break the experience. Do things you're not supposed to. Click buttons wildly, navigate to the wrong screen, or enter invalid answers to see what happens.
- Look for visual inconsistencies or mistakes, like improper font families or sizes, off-palette colors, branding issues, or poor alignment. Your style guide is valuable here.
- Find areas where the user flow seems jarring or awkward. This isn't a bug check, exactly. You want to see if the narrative feels natural once everything's in place.
- Keep an eye out for spelling and grammar issues. Hopefully, you spell-checked and asked others to proofread your writing, but pobody's nerfect! Typos happen.
- Test your navigation and progress controls, and confirm links behave as expected. Are there any issues as you move between slides? Can you get to the end?
- Check your guardrails. In other words: can users do things they're not supposed to do? Suppose a tester clicks the wrong answer in a quiz but proceeds to the next slide without feedback or correction. Something's wrong under the hood!
- When using images, sound, or video, scan for media-related hiccups. Look for optimization problems like slow loading times, stuttering playback, or visuals that appear blurry, pixelated, or over-compressed.

- Review your requirements to see which platforms are in scope, then test on relevant browsers, devices, and operating systems.
- Capture all issues in a bug or task-tracking system. Ensure your team can see what's been fixed and what's still active.

There's plenty more where those came from, but these should get you started. I just have a few lingering suggestions before we wrap this chapter.

QA pitfalls and booby traps!

Did that get your attention? Good. Because this one's important.

The last thing I'd do is discourage informal testing. Time and budget are always limited, and tackling QA with a small, scrappy team is empowering. But some mistakes can have performance and legal implications if you're not careful.

I'll make this super clear: I offer no legal advice, and nothing in this book should be taken as such. My only goal is to raise issues so you can discuss them with your team and client, then decide how you'd like to proceed. I leave it to you to resolve what's needed for your project.

Ominous disclaimers aside, here are some topics I strongly recommend you look into.

Accessibility

Legal measures exist to ensure the web is accessible to everyone, including those with disabilities. Online requirements can be complex and include things like font sizes, visual contrast, screen reader compatibility, code formatting, and the usability of your forms and UI.

Ignoring these guidelines may cause all kinds of trouble. Addressing them at the last minute can be stressful and expensive. My advice is to discuss accessibility with your client early, work with them to lock down requirements, then write a test plan to ensure you actually meet those requirements. Emphasizing that last point: *defining* your accessibility goals is not the same as *achieving* them in your live experience. Hence, the importance of testing.

Concerning QA, the levels of compliance can be nuanced, and it's tricky to mimic assistive equipment (e.g., screen readers). For these reasons, companies often hire third parties who specialize in accessibility testing.

If you're leveraging an LMS or pre-built framework, you may have limited control under the hood. But accessibility is always worth a discussion, even if you and your client decide it's irrelevant to your situation. Although, personally, I'd get that in writing.

Please don't rely on this blurb as your source of truth. There's a lot more to it, and standards constantly evolve. Some tools can even help evaluate your comps and code for everything we discussed! If any of this relates to your project, I recommend looking up the following for details:

- Web Content Accessibility Guidelines (WCAG)
- American Disabilities Act (ADA)
- Section 508

I know this may sound intimidating, but it's worth exploring before you go live. Do your research. Help your client define requirements. Ignore accessibility at your peril.

As usual, I've added a few starters to *Tools & Resources*.

Performance

Everyone wants their experience to download faster, be more responsive, and play media without stuttering. Unfortunately, loading and performance issues can have any number of causes.

While we can't cover them all, here are some typical culprits when dealing with such things. Some may be easier to fix than others.

- **Media optimization**: If images or videos aren't properly optimized for web, you can have problems with slow downloads and playback. Production design is an art and science, and finding a balance between good looks and fast downloads is part of the job. Encoding video is especially dodgy, so if you're running into media problems, you'll need to do some research or track down someone who knows their stuff. At a minimum, it's worth educating yourself on reasonable image file sizes for online use. Prepping images is simpler than exporting video, and you want to avoid uploading huge bitmaps unless your system automatically optimizes them.
- **Code efficiency**: Like media, custom code should be optimized to prevent performance issues. We won't linger here, but excessive file sizes and unnecessary loops can affect responsiveness for your users. Modern libraries and frameworks can help as well.

- **Hosting and content distribution:** No matter how well-optimized your source is, the speed at which code and media arrive is limited by the system that serves them. Your *content distribution network* (CDN) or hosting provider sometimes contributes to slower downloads.

Cross-browser and cross-platform issues

As mentioned in the Bug Hunter list, ensure you can test (or emulate) any browsers and platforms from your client's requirements. Also, it's prudent to specify these details in your statement of work—hardware, software, and oldest supported versions. Leave these out, and your client may assume your eLearning functions everywhere. You don't want to build for antique tech just because your stakeholders refuse to upgrade!

Remember that every configuration requires time to test and remediate. Fair warning: not everything works the same on all browsers or devices, especially the older ones.

Localization and internationalization

I won't even try to tackle *localization* (translating language) and *internationalization* (customizing content for various locales). Although, I've included some links at the back if you'd like a primer.

Suffice it to say, when recreating your content for multiple audiences, make sure you're prepared to write, enter, and test everything accordingly. You'll need to build your experience to handle separate, translated versions.

Also, guess what? Most typefaces can't handle every language, and not every language flows in the same direction! All this can significantly impact your design and budget.

Don't freak out

I know. That's a lot to consider in addition to the work you're already doing. But my goal here is to share all the puzzle pieces. Talk through requirements with your team and client. Decide what's necessary for your audience. Take it one step at a time.

If these issues apply, you may want to add them to your documentation. And if you need specialists to check accessibility or performance, settle that early so you can gauge the impact.

Chapter takeaways

Nicely done! You're nearing the end of the road. We talked about building your live eLearning experience, ensuring everything works as expected, and critical considerations.

- It's time to apply the answers you gathered in the *Get organized and choose your tools* chapter. You should know who's handling technical development, what methods or authoring tools you'll use, and sources for media assets.
- In this step, you'll create live production assets and build the experience based on requirements.
- There are several methods for testing an eLearning experience. Some projects recruit small, informal teams, while others enlist dedicated staff.
- *Quality assurance (QA) testing* identifies bugs and ensures an experience aligns with the original requirements.

- *User acceptance testing (UAT)* considers a real-world user's perspective.
- Know how to look for bugs and common issues. Test again after fixing those issues.
- When applicable, plan for accessibility, cross-browser and cross-platform concerns, localization, and internationalization. Discuss these requirements early to avoid last-minute problems.

19

IMPLEMENT AND EVALUATE

Whew! You've designed and built your eLearning experience, adapted to client feedback, and tested the heck out of it. Time to roll it out to your learners and see how it fares in the wild.

If you've been prototyping along the way, this may be less Grand Reveal and more What You've Been Doing All Along, Now With Polished Content! Still, you can only predict so much during development. You'll get more insights from real-world use.

In this chapter, we'll discuss ways to evaluate your training. We'll find tweaks to improve your experience and see which activities resonate with learners.

Implementation

Unfortunately, there's no universal Launch Now button to deploy an eLearning experience. You built your training on specific systems, media, and technology; those choices will dictate how you go live (or *Implement* in ADDIE lingo).

If you're using an LMS, this process will look very different from developing from scratch. Learning management systems save the hassle and maintenance of building custom tools, but there's still plenty of prep before launch. You may need to upload course files and configure the system with profiles, time limits, or thresholds for a passing grade. Options vary by LMS.

If you've gone the custom development route, you'll need to do all that heavy lifting yourself. You may want to write course management plans and decide how to gather analytics. Pushing code live is entirely on you and your devs, so hopefully, you've chosen your tech stack with care.

Regardless of your infrastructure, the goal is to share your experience with your audience and organize learning sessions. And before you're up and running, you'll want an evaluation plan ready to go.

Evaluation

The goal of *evaluation* is to see how things are working and find places to improve. This might mean gathering survey feedback, analyzing performance from activities, or interviewing learners directly. But how do you structure a project-wide overview?

We'll explore industry methods shortly, but in most cases, you'll do the following.

1. Draft an evaluation plan with questions, objectives, and timelines.
2. Define a sample size and ways to collect information: interviews, surveys, etc.
3. Set up your data collection methods. If you're using an LMS, chances are you'll have access to built-in

reporting tools. Otherwise, you'll need to prepare them yourself.

4. Collect data as people use your eLearning. Sampling at different stages can help you track progress and find areas for improvement.
5. Study your data and look for insights. Statistical analysis is helpful here, and your LMS may offer tools to parse, sort, and organize information.
6. Think about changes you'd make based on your findings. Then work with developers to assess the levels of effort.
7. Review your results with stakeholders and propose the next steps.
8. Update your eLearning as needed. Much like the QA process, you'll want to reevaluate after each change to confirm your tweaks are going in the right direction.

This might sound like high-tech space magic, and it can be. But so is AI. And airplanes. And memory foam! And just like those things, digital surveys and analytics have become pretty accessible. But even with the right tool, you still need a strategy.

With that in mind, let's look at a proven model.

The Kirkpatrick Model

The *Kirkpatrick Model of Evaluation* is used to assess the effectiveness of training. Donald Kirkpatrick proposed the idea in 1959, and it's still common in both corporate and academic settings. The framework outlines four levels of evaluation: *reaction*, *learning*, *behavior*, and *results* (Kirkpatrick & Kirkpatrick, 2006).

- *Reaction* refers to participants' immediate responses to a training program. Did they like it? Did they find it helpful?
- *Learning* refers to participants' understanding of the material and their ability to apply it. Did they gain new knowledge or skills?
- *Behavior* refers to their actual conduct after the training. Did they use what they learned?
- *Results* refer to the impact of the training on participants' work performance or other outcomes. Did it improve their job performance or help them achieve other goals?

To use the model, you'll focus on one level of evaluation and collect data accordingly. For example, to gauge participants' opinions about training (the reaction level), you might survey them immediately after a session. When assessing their ability to apply knowledge (the learning level), you might add a test at the end. And so on.

You can get creative with data collection tactics (and we'll discuss that, too), but that's the idea. Move through each level to gather insights, then decide whether to modify or adapt your training in that area.

Rather than just asking if your training worked, the Kirkpatrick Model guides your review through different lenses. This makes it easier to organize your results and share outcomes with stakeholders.

When should you evaluate?

Evaluations come in all shapes, but the options for *when* to evaluate fall into three main categories.

- A *summative evaluation* is conducted at the end of a project, program, or period to assess overall effectiveness. This approach asks a broad question: "Did your training work as intended?" If the answer is no, you'll need to look deeper and examine what went wrong, where, and how to fix the issue.
- *Formative evaluations* run continuously. They provide feedback to improve your training while it's ongoing.
- Finally, *confirmative evaluations* verify that a program is meeting its objectives for the long haul. Some consider this approach an extension of the other two because it uses similar methods but adds more time. Or rather, it asks: "Are your learning materials *still* effective?" These evaluations measure results weeks, months, or even years after the initial training.

In practice, you may combine these methods a bit.

Summative evaluations help you gauge overall success. Formative evaluations let you tweak in real time if you're looking to adjust training in progress. And to see whether long-term projects stay relevant, confirmative evaluations may be worth the investment.

Whatever the case, you'll need ways to gather data!

Creating surveys and evaluation forms

A chunk of your insights will come directly from learner feedback. If you're looking for thoughts on satisfaction and impact, it's best to tap into the source, right?

But you'll need a means of collecting feedback before you can analyze it. That means coming up with questions and deciding how to present them. Surveys, focus groups, or interviews are all helpful, depending on the situation.

Since we're focusing on eLearning, let's take a look at digital surveys. They're asynchronous and can be shared with a large workforce. A search for "online survey tools" should yield plenty of options, or you can flip to *Tools & Resources* for inspiration.

Designing your surveys

Digital surveys use code-based *forms* to collect responses for each question. You might offer multiple choices or ask learners to write answers essay-style. You could implement checkboxes, drop-down menus, or long-form text. Each form element has characteristics that affect your survey design.

- **Interface:** What does the UI look like, and how does it work across devices? A drop-down list (called a *select* in HTML) may function differently between your laptop and phone.
- **Behavior:** Do you want users to choose only one option, select multiple options, or enter unlimited text?
- **Depth:** How much information do you need in response to a question? Do you want a single word or phrase, or are you looking for deeper feedback? You won't get freeform insights from a multiple-choice question, but maybe all you need is a yes or no.

Start by collecting questions that spring to mind. From training impressions to activities your learners found challenging. Arrange these in a progression and have one idea flow into the next.

Review your questions and think about how to present them. Some will work fine with short (or even binary) answers. For others, you may want detailed responses.

With each question in your survey, consider:

- What do I want my input UI to look like?
- How many options will I offer?
- How important is this question?
- What level of detail would help?
- Am I looking for unexpected answers or simple responses?
- Should this be optional or required?

You can then start assembling your form. Does a question lend itself to:

- Radio buttons (one selection)
- Checkboxes (multiple selections)
- Short text fields (one line)
- Long-form text fields (for elaboration)
- Drop-downs (allow for many options in a compact space)
- Variations of the above, like star ratings or sliders (often built with radio buttons under the hood)

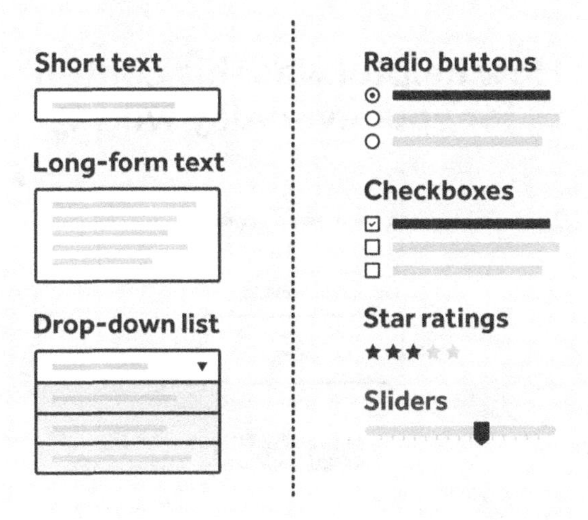

For each question, match the interface to your feedback needs. For example, to gauge satisfaction, you could use a radio-button rating system (e.g., On a scale of "meh" to "outstanding," how helpful was this content for your everyday work?), followed by a text field for elaboration. Maybe the rating is required while details are optional, helping you collect feedback without annoying less-invested users.

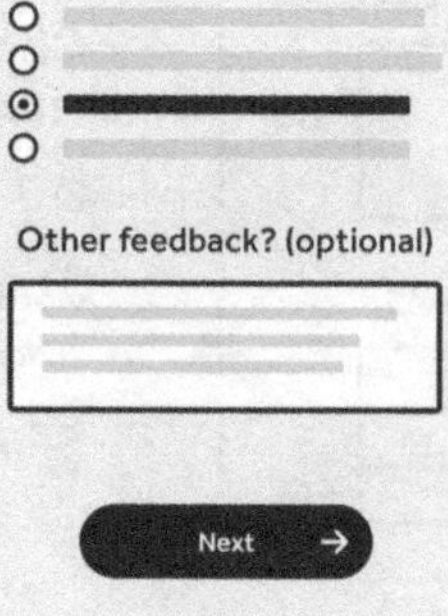

You'll also want to consider spatial issues. A long list of radio buttons can be unwieldy with more than a handful of options —when a user scrolls to the bottom, your original question might be off-screen! It may be better to collapse answers into a drop-down menu, especially if you have several questions per page.

Dig beneath the surface

Shallow questions like "Did you enjoy the material?" aren't verboten, but they offer limited insight and aren't very actionable. Sure, they'll give you a vague sense of learner satisfaction. Beyond that, how would you use those responses?

Write questions to gauge the quality of your content, its relevance to learners' day-to-day lives, and the value of your resources. Poke around for challenging spots and unclear messaging.

Ask about specifics, like:

- **Media:** Did learners find animations, videos, and other media helpful?
- **Development:** Was your eLearning well-built? Or did it suffer from performance issues or glitches, like answers not being saved?
- **Time:** Did learners have enough time to finish the course without rushing? Was the training too long?
- **Usefulness:** Did learners feel the content was useful in practice? What would make it more so? What activities did they find most and least applicable?

In the spirit of applying what you've learned: how would you reframe or present these questions in survey form? Which might work with simple ratings, and where would you leave room for long answers?

Analyzing results

Piles of data are only useful when you decipher the results. And although the brain is one of my all-time favorite organs, such tasks may be hard to do manually. Fortunately, some eLearning tools can help.

In addition to serving as a survey-creation hub, your LMS may offer deep analysis of your learners—average completion times, scores by user group, or areas of common confusion. Features vary, so be sure to read up on your particular system.

Ideally, your analysis tools produce human-readable reports, helping you make informed decisions and improve your work.

Chapter takeaways

Implementation is a major step in your project, but you ain't done just yet! After your eLearning goes live, you still need to gauge its effectiveness. You'll decide how to collect learner data, structure your evaluations, and design surveys to gather insights and feedback.

- Your steps for *implementation* (pushing your eLearning live) will vary with your tech approach. Deployment for an experience built with an LMS will differ from one developed from scratch.
- You'll create a plan to *evaluate* live training and define your approach, tactics, and timeline.
- *The Kirkpatrick Model* is a long-established system for instructional designers. It examines four levels of the learner experience: *reaction*, *learning*, *behavior*, and *results*.
- *Summative*, *formative*, and *confirmative evaluations* are different sampling methods that help you consider when to evaluate and how often.
- Designing good surveys requires a creative touch. Ensure the user interface aligns with each question and probe for actionable feedback.
- Deciphering evaluation data can be complex. Digital tools from your LMS or other systems can make this easier and help you adapt.

20

END OF THE ROAD

Well done! Whatever your reasons for reading this book, hopefully you've gained a firm foundation and some new tricks for your bag.

You learned the nuts and bolts of instructional design, explored models, principles, and frameworks, and found ways to apply them in a practical workflow. As you continue your professional journey, remember that—beneath all the process and jargon—these tools are meant to build something. To leverage human patterns and help people learn.

So go ahead and try these things out. See how each method performs in your projects and hold nothing precious. Keep what works for you, and design your own roadmap!

A shameless, parting plug

However you proceed in your career or academic pursuits, I hope you found this book helpful. Like you, I've sifted through reams of data to decipher this practical-but-deep field. I've worked hard to slice all these dense cuts into chewable bits.

If you liked what you read, consider leaving an Amazon review. Your support helps me continue this work and is, of course, hugely appreciated.

Thanks for hanging with me, and safe travels!

TOOLS & RESOURCES

At the following link, you'll find example files from the book, software starters, and utilities for design, QA, and accessibility.

While I offer this page as guidance, I suggest flexibility. Apps grow outdated or lose favor. Standards change. New AI tools emerge almost hourly.

Keep your eyes peeled for innovation, then swap these resources as you find your own.

Enjoy!

www.aubrey-cook.com/resources
password: **love2cook**

COLLECTED QUESTIONS

Below, you'll find all the questions I've scattered throughout this book. May they aid your design work, client discussions, and team meetings!

CHAPTER 5: Get organized and choose your tools

Figure out who you're working with

- Who knows the most about current issues and goals?
- Who can help define the business objectives?
- Who has expertise in existing material or relevant topics?
- Who can help me understand the target audience?
- Who will make final decisions?
- Who will provide feedback?

How will you build your eLearning experience?

- Will your eLearning involve custom development?
- Do you have access to someone (yourself, a team, or a specialized resource) who can build whatever you design?
- Are you or your client bound by an existing learning management system (LMS)?
- Is it part of your job to decide the best platform for your client, starting from scratch?
- Will you use eLearning authoring software to create lessons and interactive material?
- Are there features that might be challenging with your chosen technology?
- How will you track and analyze learners' performance?
- What digital devices or platforms will you support?
- What are your accessibility requirements?
- Should you plan for multiple languages or regional content?

How will you create or acquire multimedia?

- What types of visual assets will you use in your layouts, storyboards, and final learning experience?
- How will you create or acquire video, animation, photography, illustrations, or iconography? Are you producing them yourself or paying for stock media?
- Does your eLearning authoring tool (if you're using one) offer character or illustration libraries?
- Does your client already have existing assets they want you to use, or will everything need to be custom?
- Do they have the budget to purchase or create whatever assets are required?

CHAPTER 6: A taste of project management

Scoping and estimation

- How many rounds of work will your client want for each deliverable?
- What is their internal review process, and how many people does it involve?
- How long will your client need for each internal review?

Milestone placement

- Where might stakeholder feedback have the biggest impact?
- Are there long work blocks where a misstep could force you to backtrack?
- When will you need all your stakeholders to gather and agree on something?

CHAPTER 8: Work with SMEs and stakeholders

Distinguish impactful vs. extraneous material

- How and when would a learner use this in practice?
- What would go wrong if learners didn't know this?
- If every learner had this information, how does it impact the business goal?

Fish for scenarios

- What problems have you seen, and what were the circumstances?
- What happened as a result of mistakes?
- Were there consequences or attempted remedies?
- What worked, and what didn't?
- What made situations stressful?
- Were there conditions, settings, or details that set the stage?
- What would successful outcomes look like in each case, and what actions could the learner take?

Possible challenges

- Can you think of obstacles that might derail your eLearning? For example: were there issues with a lack of training time, spotty leadership support, or learners' discomfort with technology?

CHAPTER 9: Analyze your learners, needs, and tasks

Needs assessment

- Do you have existing surveys, analytics, or feedback? What data suggested an issue?
- What are employees currently doing, and what's wrong with it?
- What are the goals for your employees? How would you like them to perform?

- If they are missing specific targets, why? Look for root causes here: issues in their environment, missing skills or resources, or lackluster motivation.
- What skills and knowledge do employees currently have?
- What skills and knowledge must employees develop?
- What resources (e.g., materials, equipment, people) could help employees perform better?
- What obstacles (e.g., behavior, environment, motivation, knowledge, or skill) might get in the way?

Task analysis

- How often is this step performed?
- How difficult is it for your target audience?
- What knowledge or abilities does it require to complete?
- How crucial is this step for accomplishing the primary goal?
- What are the consequences if this step is omitted or done wrong?

Learner analysis

- Can you provide some context about learners' current knowledge and skill levels?
- Do they learn best from video webinars, interactive experiences, or live group sessions?
- What devices do they use most frequently, and what's their technological comfort level?
- Does this group prefer face-to-face instruction?

Context analysis

- Where will your training take place, and what's the environment like?
- Are there distractions, conditions, or deprecated technology you need to deal with?
- Is your eLearning remote?

CHAPTER 10: Build objectives, actions, and practice activities

- Based on your business goal, what should learners be able to do after they finish your lessons or training?
- How will you structure your material to help them get there?

Better objectives with Bloom's Taxonomy

- How will people use their new knowledge? In other words, what kinds of tasks, interpretations, or activities will they have to do?
- How deeply must people learn the material to do these things, and how much effort should they invest?

Action mapping

- What's wrong with the current situation?
- What can people do to address these things?
- What are they currently doing well?
- What could they do better? What changes could they make?
- What isn't getting done that should be?

- What are the consequences if the goal isn't reached?
- What changes would have the biggest impact?
- Which actions influence the goal most?
- What actions have the least impact?
- How can we address the most crucial actions?
- Which actions can you manage with references, checklists, or job aids? In other words, which solutions don't require in-depth training?
- Which remaining actions benefit from practice? Which require interpretation, expertise, and memorization?
- What specific activities would help learners practice what we need them to do?
- Which of these activities can you simulate with eLearning?

CHAPTER 13: Evolve the outline into a storyboard

Sourcing your media

- Will this project require video, audio, illustrations, or still images?
- Will you or your team produce assets from scratch based on any scenario you might dream up?
- Are you limited to finding and buying pre-made stock images?
- Does your eLearning software have character libraries?

CHAPTER 15: Create a style guide and layout templates

Create a style guide

- Is there an existing brand guide?
- Are there logo and branding assets we can use?
- Do you have references or guidance for visual design?
- How would you like to represent characters or people?

CHAPTER 16: Prototyping

What answers do you need from your audience?

- Which content, sections, or scenarios should you mock up?
- Do you need to demonstrate multiple paths, or can you make an on-rails linear experience?
- How much interactivity, animation, and interface detail do you need to accomplish the goal of your prototype?
- Can you build this yourself, or do you need help from others?
- Will your audience get distracted if something is low fidelity?
- What impact would higher fidelity have on your timeline and budget?

CHAPTER 18: Develop and test

Uploading content and media

- Are all your assets exported, optimized, and ready to post?
- Who will upload media and written content? And where does that happen?

Make sure everything works as intended

- What does the QA and UAT process entail, and what do you need to plan for in the project plan?
- Will you be...
- ...testing yourself?
- ...recruiting coworkers to help find issues, especially those unfamiliar with the work?
- ...sending the experience to a dedicated QA team for full review and bug reporting?
- ...engaging third parties to assess optimization, performance, or accessibility compliance?

REFERENCES

Allen, M. W., & Sites, R. (2012). *Leaving ADDIE for SAM: An agile model for developing the best learning experiences*. Association for Talent Development.

Anderson, L. W., & Krathwohl, D. R. (Eds.). (2001). *A taxonomy for learning, teaching, and assessing: A revision of Bloom's taxonomy of educational objectives* (Complete ed). Longman.

Armstrong, P. (2010). *Bloom's Taxonomy*. Vanderbilt University Center for Teaching. Retrieved February 28, 2023, from https://cft.vanderbilt.edu/guides-sub-pages/blooms-taxonomy/

Bloom, B. S. (1984). *Taxonomy of Educational Objectives, Handbook 1: Cognitive Domain* (2nd ed.). Addison-Wesley Longman Ltd.

Branson, R. K., Rayner, G. T., Cox, J. L., Furman, J. P., King, F. J., & Hannum, W. H. (1975). *Interservice procedures for instructional systems development: Executive Summary and Model.* Center for Educational Technology, Florida State University.

Clark, R. C., & Mayer, R. E. (2016). *E-learning and the science of instruction: Proven guidelines for consumers and designers of multimedia learning* (Fourth edition). Wiley.

Dick, W., & Carey, L. (1978). *The Systematic Design of Instruction*. Scott, Foresman.

Floor, N. (2021, January 21). *Learning Experience Design vs Instructional Design —LXD.org blog*. Learning Experience Design. https://lxd.org/news/learning-experience-design-vs-instructional-design/

Gagné, R. M. (1985). *The conditions of learning and theory of instruction* (4th ed). Holt, Rinehart and Winston.

Kirkpatrick, D. L., & Kirkpatrick, J. D. (2006). *Evaluating training programs: The four levels* (3rd ed). Berrett-Koehler.

Knowles, M. S. (Ed.). (1984). *Andragogy in action* (1st ed). Jossey-Bass.

Lowe, R. K. (2004). Animation and learning: Value for money? In R. Phillips, D. Jonas-Dwyer, C. McBeath, & R. Atkinson (Eds.), *Beyond the Comfort Zone: Proceedings of the 21st ASCILITE Conference* (pp. 558–561).

Merrill, M. D. (2002). First Principles of Instruction. *Educational Technology Research and Development*, *50*(3), 43–59.

Moore, C. (n.d.). *Will training help?* Retrieved June 20, 2023, from https://blog.cathy-moore.com/will-training-help/

Moore, C. (2017). *Map It: The hands-on guide to strategic training design*. Montesa Press.

Smith, M. K. (2002). *Malcolm Knowles, informal adult education, self-direction and andragogy*. The Encyclopedia of Pedagogy and Informal Education. https://infed.org/mobi/malcolm-knowles-informal-adult-education-self-direction-and-andragogy/

KEEP READING

Oh, you thought we were done? We're never done!

There's always more to learn in this field, and there are so many great minds. I've included work from oldies but goodies, wisdom from modern pros, and links I found helpful for untangling the brain.

Time and the internet will erode this list until it's a ghost of the original, but by then, we may all be uploaded to the cloud anyway.

Get 'em while you can, and I'll see you all up there!

Atrash, D. (2022, February 8). *Understanding Web Accessibility Standards: ADA, Section 508, and WCAG compliance*. Medium. https://bootcamp.uxdesign.cc/understanding-web-accessibility-standards-ada-section-508-and-wcag-compliance-143cfb8b691e

Babich, N. (2016, March 2). *Responsive Design Best Practices*. Medium. https://uxplanet.org/responsive-design-best-practices-c6d3f5fd163b

Boller, S., & Fletcher, L. (2020). *Design thinking for training and development: Creating learning journeys that get results*. ATD Press.

Bonneville, D. (2010, November 4). *Best Practices Of Combining Typefaces*. Smashing Magazine. https://www.smashingmagazine.com/2010/11/best-practices-of-combining-typefaces/

Brown, T. (2016, April 29). Combining Typefaces: Free guide to great typography. *The Typekit Blog*. https://blog.typekit.com/2016/04/29/combining-typefaces-free-guide-to-great-typography/

Da Silva, M. (2022, January 24). *ADDIE: The Instructional Designers' Best Friend*. ELearning Industry. https://elearningindustry.com/addie-the-instructional-designers-best-friend

DeBell, A. (2019, December 11). How to Use Mayer's 12 Principles of Multimedia Learning. *Water Bear Learning*. https://waterbearlearning.com/mayers-principles-multimedia-learning/

Dirksen, J. (2016). *Design for how people learn* (2nd ed.). New Riders.

Elkins, D. (2011, May 5). *Negotiating Out "Nice-to-Know" Information*. E-Learning Uncovered. https://elearninguncovered.com/2011/05/negotiating-out-%e2%80%9cnice-to-know%e2%80%9d-information/

Fonts Knowledge. (n.d.). Google Fonts. Retrieved June 22, 2023, from https://fonts.google.com/knowledge

Friedman, V. (2018, August 11). *Responsive Web Design: What It Is And How To Use It*. Smashing Magazine. https://www.smashingmagazine.com/2011/01/guidelines-for-responsive-web-design/

Guidance on Web Accessibility and the ADA. (2022, March 18). ADA.Gov. https://www.ada.gov/resources/web-guidance/

Hellmuth, M. (2020, January 11). *Everything you need to know about spacing & layout grids*. https://www.uiprep.com/blog/everything-you-need-to-know-about-spacing-layout-grids

Herrholtz, K. (2020, March 6). *Rapid Instructional Design With SAM*. ELearning Industry. https://elearningindustry.com/sam-successive-approximation-model-for-rapid-instructional-design

Initiative (WAI), W. W. A. (n.d.). *Accessibility Fundamentals Overview*. Web Accessibility Initiative (WAI). Retrieved June 22, 2023, from https://www.w3.org/WAI/fundamentals/

Ishida, R., & Miller, S. K. (2005, December 5). Localization vs. Internationalization. W3C. https://www.w3.org/International/questions/qa-i18n

Khagwal, N. (2020, November 23). *Responsive Grid Design: Ultimate Guide*. Medium. https://medium.muz.li/responsive-grid-design-ultimate-guide-7aa41ca7892

Kliever, J. (n.d.). *A beautifully illustrated glossary of typographic terms you should know*. Learn. Retrieved February 28, 2023, from https://www.canva.com/learn/typography-terms/

Kurt, Dr. S. (2021, February 17). Instructional Design Models and Theories. *Educational Technology*. https://educationaltechnology.net/instructional-design-models-and-theories/

LePage, P., & Andrew, R. (2022, September 22). *Responsive web design basics*. Web.Dev. https://web.dev/responsive-web-design-basics/

Lupton, E. (2010). *Thinking with type: A critical guide for designers, writers, editors, & students* (2nd ed.). Princeton Architectural Press.

Marcotte, E. (2010, May 25). *Responsive Web Design*. A List Apart. https://alistapart.com/article/responsive-web-design/

Maria, J. S. (2009, November 17). *On Web Typography*. A List Apart. https://alistapart.com/article/on-web-typography/

McDonald, J., & West, R. (2021). *Design for Learning: Principles, Processes, & Praxis* (1st ed.). EdTech Books. https://doi.org/10.59668/id

Merrill, M. D. (2007). First principles of instruction: A synthesis. In R. A. Reiser & J. V. Dempsey (Eds.), *Trends and Issues in Instructional Design and Technology* (2nd ed., Vol. 2, pp. 62–71). Merrill/Prentice Hall.

Merrill, M. D. (2009). First Principles of Instruction. In C. M. Reigeluth & A. A. Carr-Chellman (Eds.), *Instructional Design Theories and Models: Building a Common Knowledge Base* (Vol. 3). Routledge Publishers.

Moore, C. (n.d.-a). *Action mapping FAQs*. Retrieved February 28, 2023, from https://blog.cathy-moore.com/action-mapping/action-mapping-faqs/

Moore, C. (n.d.-b). *Action mapping on one page*. Retrieved February 28, 2023, from https://blog.cathy-moore.com/online-learning-conference-anti-handout/

Moore, C. (n.d.-c). *How to create a training goal in 2 quick steps*. Retrieved June 20, 2023, from https://blog.cathy-moore.com/how-to-create-a-training-goal-in-2-quick-steps/

Pappas, C. (2021, January 13). *ADDIE Model Vs SAM Model: Which Is Best For Your Next eLearning Project*. ELearning Industry. https://elearningindustry.com/addie-vs-sam-model-best-for-next-elearning-project

Peck, D. (2023a, May 4). *How to Create a Storyboard for eLearning (Instructional Design).* https://www.devlinpeck.com/content/create-storyboard-for-elearning

Peck, D. (2023b, May 5). *4 Types of Analysis for Instructional Design.* https://www.devlinpeck.com/content/analysis-instructional-design

Section508.gov. (n.d.). Retrieved June 21, 2023, from https://www.section508.gov/

Slade, T. (n.d.). How to Conduct a Needs Analysis. *The ELearning Designer's Academy*. Retrieved February 28, 2023, from https://elearningacademy.io/blog/how-to-conduct-a-needs-analysis/

Slade, T. (2020). *The eLearning Designer's Handbook: A Practical Guide to the eLearning Development Process for New eLearning Designers* (2nd ed.). Independently Published.

Spiekermann, E., & Ginger, E. M. (1993). *Stop stealing sheep & find out how type works*. Adobe Press.

Stocks, E. J. (n.d.). *Introducing weights & styles*. Google Fonts. Retrieved February 28, 2023, from https://fonts.google.com/knowledge/introducing_type/introducing_weights_styles

Strehlow, R. (2021, November 8). *Typefaces vs. fonts: Here's how they're different*. Shaping Design Blog. https://www.editorx.com/shaping-design/article/typefaces-vs-fonts

Team Designlab. (2022, August 25). *The Grid System: Importance of a Solid UX/UI Layout.* https://designlab.com/blog/grid-systems-history-ux-ui-layout/

WCAG 2 Overview. (n.d.). Web Accessibility Initiative (WAI). Retrieved June 21, 2023, from https://www.w3.org/WAI/standards-guidelines/wcag/

Weber, M. (2019, November 27). *Kirkpatrick's Evaluation Model.* ELearning Industry. https://elearningindustry.com/kirkpatricks-evaluation-model-levels

Williams, R. (2015). *The non-designer's design book: Design and typographic principles for the visual novice* (Fourth edition). Peachpit Press.

ACKNOWLEDGMENTS

My deepest thanks to Chandi Lyn for her keen edits and insights, and for tamping my humor to a family-friendly level.

I'm grateful to Marcus, Bill, Katherine, and Rex, whose revitalizing company kept my brain from turning to goop on this journey.

Above all, thanks to my parents, for championing my obsessions since birth.

Made in United States
Cleveland, OH
12 January 2025

13360215R10154